Cover Image Credit/Copyright Attribution: IIIerlok_Xolms/Shutterstock

Steinar Løve Ellefmo and Fredrik Søreide
with contributions from
Georgy Cherkashov, Cyril Juliani,
Krishna Kanta Panthi, Sergey Petukhov,
Irina Poroshina, Richard Sinding-Larsen
and Ben Snook

Quantifying the Unknown

Marine Mineral Resource Potential on the Norwegian Extended Continental Shelf

This book is made possible with support from the Norwegian University of Science and Technology (NTNU).

A note about the images reproduced in this book: many of the detailed maps and illustrations are not legible in the small format of the print and PDF editions. All of the images are accessible in larger, fully legible form via links in the image captions in the digital editions.

This book is a peer-reviewed monograph.

Contents

Preface

The prospect of deep-sea mining was initially suggested by J. L. Mero in his book *Mineral Resources of the Sea* in the mid-1960s (Mero, 1965). Mero claimed that nearly limitless supplies of cobalt, nickel and other metals could be found throughout the oceans in the form of manganese nodules, which appear as lumps of compressed sediment on the sea floor at depths of about 5000 m. Nodules were originally discovered by the HMS *Challenger* expedition in the 1870s. Several nations started to investigate these claims, especially in the Clarion-Clipperton Zone in the Pacific Ocean in the 1960s. However, these initial resource estimates of deep-sea minerals turned out to be exaggerated. This overestimate, coupled with depressed metal prices, led to the near abandonment of nodule mining by 1982.

Norway was involved in several new research initiatives, primarily through the Fridtjof Nansen Institute in the 1980s and 1990s, to reinvigorate the case for seabed mining. This culminated in the establishment of the Norwegian Deep Seabed Mining Group, a consortium of five Norwegian companies, the Fridtjof Nansen Institute and Norwegian University of Science and Technology (NTNU), to develop new business cases from 1999 to 2001 (Søreide et al., 2001). However, due to the still-depressed metal prices, no profitable business cases could be identified.

Meanwhile, in 1977, a group of marine geologists discovered the first hydrothermal vents along the Galápagos Rift, part of the global Mid-Ocean Ridge (MOR) system where new oceanic crust is generated, while diving in the DSV *ALVIN*, a research submersible operated by Woods Hole Oceanographic Institution. It was slowly understood that these vent systems constitute mineral deposits in the making, delivering copper, zinc, gold and silver to the seafloor. Interest, therefore, shifted towards hydrothermal vents as another source of metals instead of scattered nodules.

Over the past decade, a new era of interest in deep-sea minerals has begun. Rising demand and diminishing supplies on land has pushed the search for new sources in many areas. Norway has a significant section of the Mid-Atlantic Ridge within its extended continental shelf. In its exploration of this part of the ridge since the early 2000s, the University of Bergen has discovered a handful of hydrothermal vent systems.

Consequently, in 2011 NTNU together with Statoil ASA (now Equinor ASA) and Nordic Ocean Resources AS established a new research initiative to investigate the deep marine minerals and mining potential in the Norwegian sector. This book presents the results of that project.

Due to the high resource potential indicated by these studies, the Norwegian government included marine minerals in its mineral strategy for the first time in 2013 and issued draft legislation for the management of marine minerals within the Norwegian extended continental shelf in 2017 (OED, 2017). This legislation (LOV-2019-03-22-7) was announced on 22 March 2019 and came into effect on 1 July 2019. Only time will tell, however, if marine mining will develop into a new industry in the future.

Trondheim, September 2019.
Steinar Løve Ellefmo and Fredrik Søreide

Introduction

A mineral is a naturally occurring, inorganic substance with a definite chemical composition and crystal structure. There are about 3500 minerals in the Earth's crust. The minerals contain elements that humankind needs daily. We are big consumers of metals like copper, zinc, gold and iron and minerals like quartz and feldspar, to name just a few. It is likely that the demand for minerals will continue to increase, with, for example, an increasing demand for renewable energy production. Without exploitation of mineral resources, we will no longer have products like white paper, toothpaste, computers, batteries, cars, windmills, buildings, glass and the infrastructure that underpins modern civilisation.

About 71% of the Earth's surface is covered by water, and it should therefore come as no surprise that the sediments occurring on the ocean floor contain mineral deposits of potential economic value (Cronan, 1992, 1999). These include:

- Aggregates, which are nearshore deposits of detrital silicate minerals and calcium carbonate, principally used in the construction industry.
- Placers, which are detrital metallic deposits generally found on beaches and in nearshore areas.
- Phosphorites, which are mineral precipitates formed in situ on the sea floor, in generally deeper water than placers.
- Manganese nodules, which are concretionary deposits principally of iron, manganese, nickel, copper and cobalt, which occur typically in the deeper parts of the oceans.
- Ferromanganese oxide crusts, or the so-called cobalt rich crusts, which occupy seamount areas in and around island chains.
- Hydrothermal deposits, which are precipitates of principally iron, copper and zinc sulphides, formed as a result of submarine volcanic activity and hydrothermal venting, which can also contain enrichments of lead, silver and gold.

The processes of supply of elements to the sea floor will determine the occurrence of these deposits. Chemical elements enter seawater in a number of different ways. Sometimes their ultimate source is the dissolved material in rivers draining the continents. In other cases, they are derived from hydrothermal solutions entering seawater from submarine volcanoes. Because ocean mixing processes are slow, elements introduced into seawater from a specific source can sometimes be identified at considerable distances from that source. Thus, the oceans are by no means uniform in composition (Cronan, 1992).

Only about 10% of the ocean floor has been mapped at a resolution greater than 100 metres. Most mineral deposits, and many ocean floor features that can indicate such deposits, cannot be seen at this resolution. We need better, smarter and more efficient technologies and data-management methodologies and systems if we want to know more about the ocean floor and its resource potential.

Marine mineral deposits are present in many areas of the ocean, some in international waters and some in exclusive economic zones (EEZ). General recognition of exclusive economic zones as potential locations for marine mineral resources other than oil and gas dates from the early 1980s. Norway sent a proposal for the extension of its continental shelf to the Commission on the Limits of the Continental Shelf (CLCS) in 2006, which was found largely acceptable in 2009. However, Norway has not issued a final proclamation of the extended boundary because of potential overlaps with other countries. The current Norwegian jurisdiction is shown in Figure 1–1, including the waters surrounding the islands of Jan Mayen and Spitsbergen.

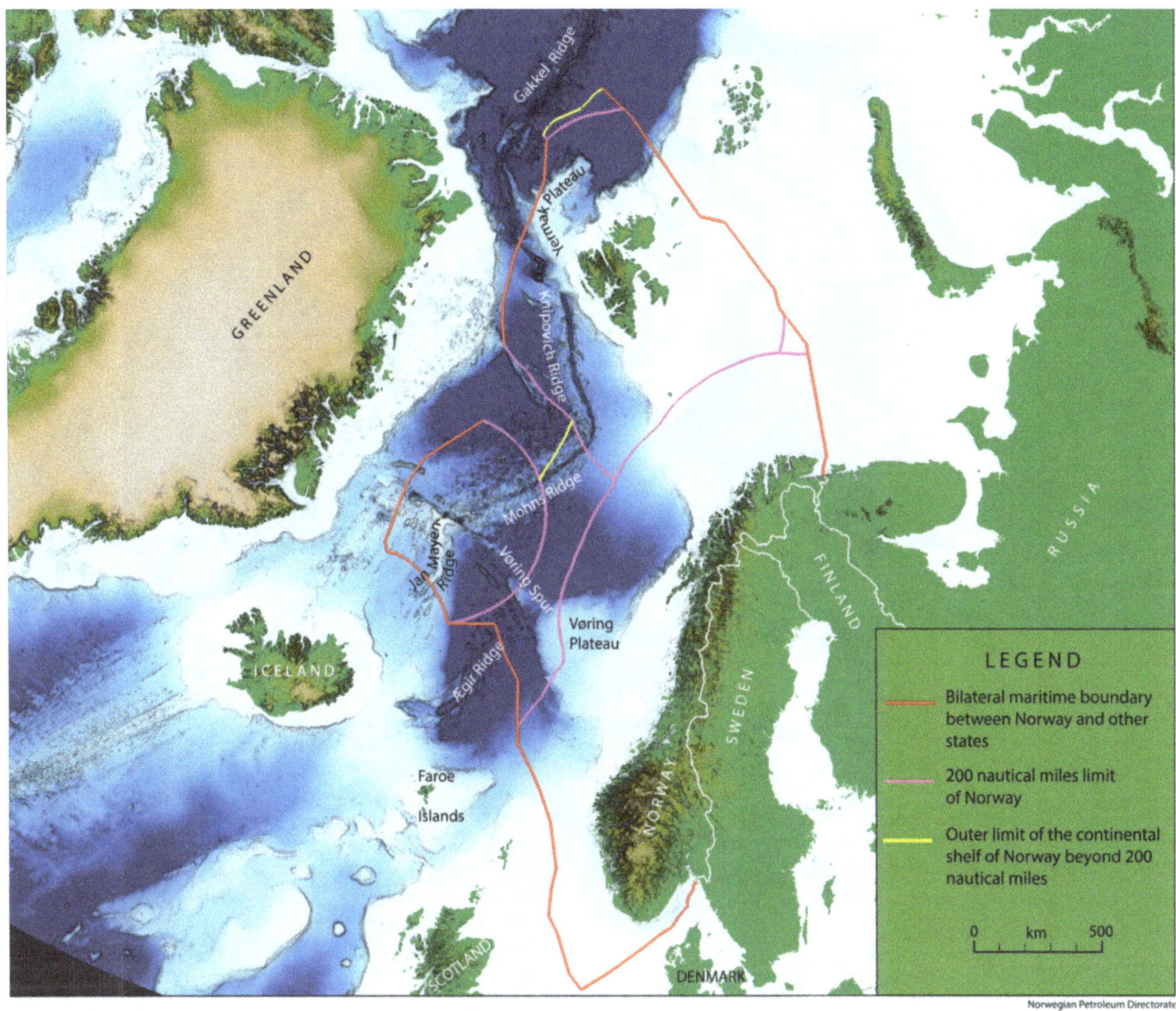

Figure 1-1 Norway manages one of the largest maritime zones in the world. The current Norwegian jurisdiction is defined by the red, yellow and purple lines on the map. Map source: Norwegian Petroleum Directorate / Harald Brekke and Becker et al. (2009). Click here for larger image.

It is possible that all of the marine mineral resources mentioned above are present to a certain extent within the Norwegian jurisdiction. However, to what extent needs to be quantified. Aggregates, hydrothermal deposits, crust and nodules have been positively identified. Figure 1–2 shows areas with potential for hydrothermal (sulphide) and crust deposits.

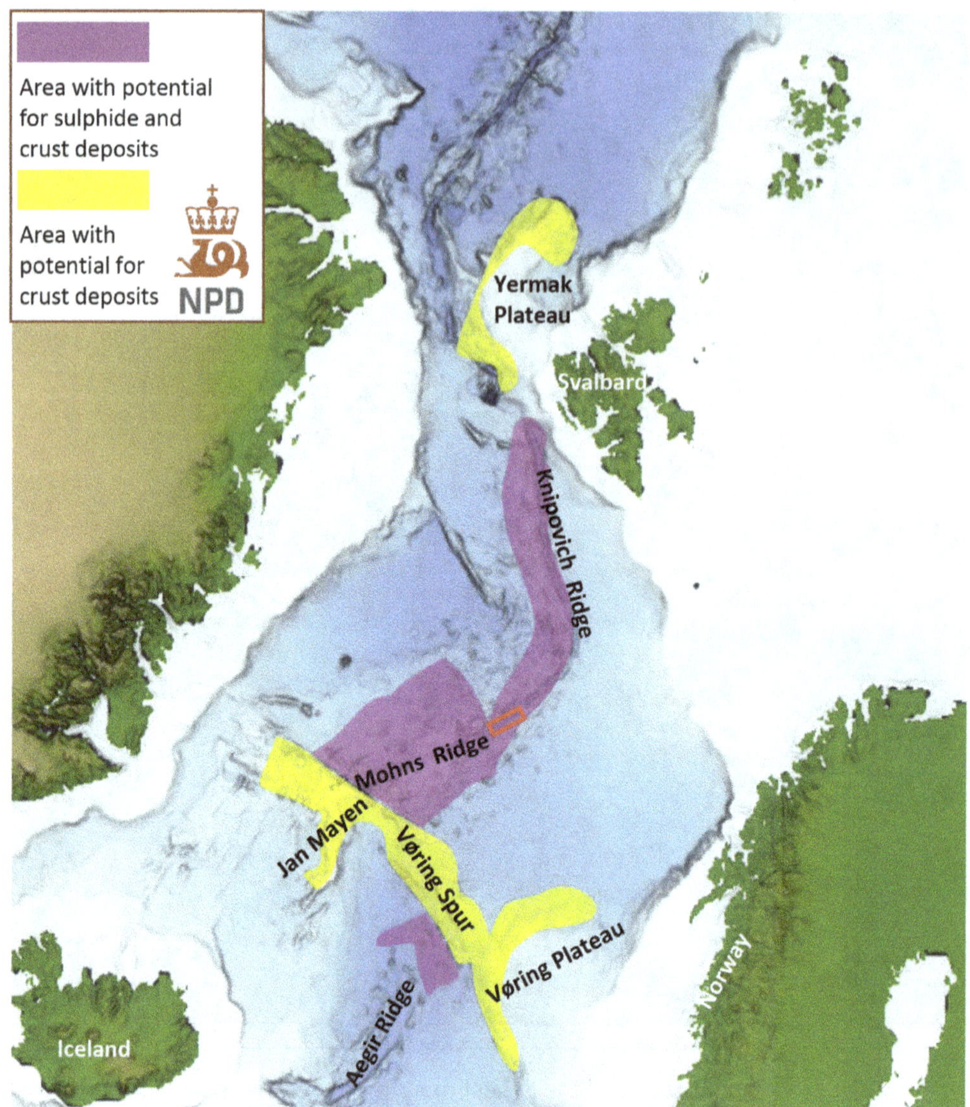

Figure 1-2 Map showing areas with a potential for sulphide and crust deposits inside Norwegian jurisdiction. The exploration area that was thoroughly investigated by the NPD in 2018 (NPD, 2018c) is indicated by the red rectangle. Map source: NPD, 2018c. Click here for larger image.

1.1 Nodules and crust

Nodules and crusts are concretions formed as a result of precipitation from the surrounding waters. Removal of elements from seawater and from the interstitial waters of sediments takes place through a number of reactions. These include direct precipitation, catalytic precipitation and absorption. The processes of direct removal of elements from sea-water into sediments are termed hydrogenous phases, and generally tend to accumulate slowly. Manganese nodules and iron and manganese-rich (FeMn) crusts are primarily hydrogenous in origin and often cobalt rich (Cronan, 1992).

Exploration for nodules has been ongoing since the 1960s. Since nodules are primarily located in international waters, this is mainly handled by the International Seabed Authority (ISA). Interest in nodule mining has seen a revival in recent years with several new exploration licenses issued. The ISA has, to date, issued 16 exploration licenses as can be seen in Figure 1–3, 15 of which are located within the so-called Clarion-Clipperton Zone in the Pacific Ocean (ISA, 2018).

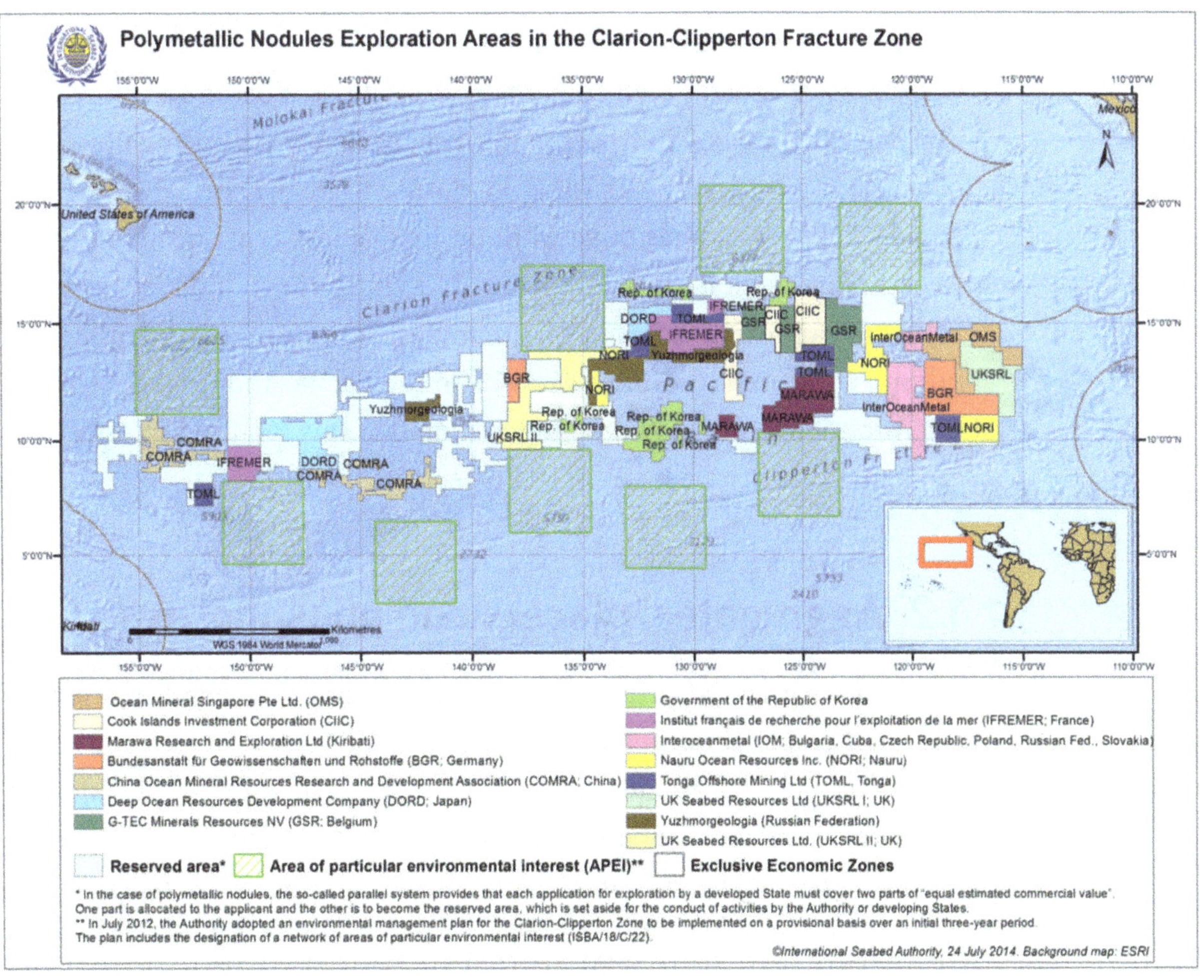

Figure 1-3 15 out of 16 exploration licenses for nodules issued by ISA are in the Clarion-Clipperton Zone in the Pacific (Image: © International Seabed Authority). Click here for larger image.

The Cook Islands also established a legal framework to issue nodule mining licenses in their EEZ in 2016. It is expected that they will begin to issue the first exploration license in 2019 (Cook Islands Seabed Minerals Authority, 2018). Four exploration licenses for FeMn crusts in the Western Pacific Ocean have currently been awarded by the International Seabed Authority (ISA, 2018).

Little is known about nodules and crusts in Norwegian jurisdiction, but findings from a 1980 cruise to ten locations in the Barents Sea indicate that ferromanganese concretions exist around Svalbard (Ingri, 1985). The deposits occur as discoidal and flat concretions and as coatings, in the latter case on lithified or detrital material or as extensive pavements on the Svalbard shelf. FeMn-rich coatings showed enrichment of Mo, Zn and Co in a Mn-oxide/oxyhydroxide mineral phase. The Fe-oxide/oxyhydroxide mineral phase holds high concretions of P and As. Two iron-rich concretions with high contents of P, Ca, St, Y, Yb and La were found east and northeast of Spitsbergen Banken, probably indicating upwelling of nutrient-rich, cold polar water along the Svalbard shelf. NPD (NPD, 2018d; Brekke, 2017) has done work on manganese crust inside the Norwegian jurisdiction. Initial chemical investigations indicate elevated values of rare earth elements (lanthanides, scandium and yttrium) and lithium compared to similar deposits in the Pacific and further south in the Atlantic Ocean. However, much more work is required to assess the economic value and density of these deposits.

1.2 Aggregates, placers and phosphorites

Terrigenous materials are the solid products of continental weathering and their composition reflects that of the continental rocks from which they are derived. They enter the oceans in a particulate form and remain in that form throughout their depositional processes. Clay minerals, quartz and feldspars are the most important silicate components in marine sediments (Cronan, 1992). These form a number of different deposit types.

Aggregates are non-metallic deposits consisting of sand, gravel, shell or coral debris and have been concentrated in their present occurrences by sea floor hydrodynamic processes, although their deposition may originally have been by some other mechanism, by rivers or glaciers, for example (Cronan, 1992). Marine aggregates are used in several countries as construction materials, including Norway.

Placers are metallic minerals or gems which have been transported to their sites of deposition in the form of solid particles and are therefore detrital minerals. They are resistant minerals which have been made available on breakdown of their parent rock. Placer diamonds are mined off the coast of Namibia by DeBeers and have become one of the main sources of diamonds.

Phosphorite is a mixed phosphate carbonate deposit, the principal mineral of which is a variety of apatite called carbonate fluorapatite, or francolite. This mineral often occurs in the deposits in the form of pellets or nodules, and at less than 1000 metres water depth. A principal requirement is a supply of phosphorous to the sea floor, often supported by marine organisms. Marine phosphorite deposits are primarily used for fertiliser production. Some deposits have been mined offshore and major new projects are currently under consideration in New Zealand and the Gulf of Mexico.

1.3 Hydrothermal deposits

Despite being located on what qualifies as an ultraslow spreading ridge, where the tectonic plates that make up the sea floor move apart at a full spreading rate of less than 2 cm per year, the environment of the mid-Atlantic ridge inside Norwegian jurisdiction is quite dynamic. This ridge section is part of the Arctic Mid-Ocean Ridge (AMOR). New crust is being constantly produced as the ocean floor creeps apart to expose the mantle, placing huge stresses on the rock, which causes extensive cracking, slipping and deformation. In addition, the spreading of the tectonic plates at the MOR causes the mantle to melt, releasing magma into the crust where some of it erupts to form volcanoes and lava flows. This leads to the formation of a ridge where faulting disrupts the topography of the surface, resulting in the formation of valleys and mountains adjacent to the main spreading margin. An extensive fracture network is also generated at depth, linked to the enormous heat source of the underlying mantle.

This system of interlocking fractures within the oceanic crust allows cold, dense seawater to migrate down through the hot rock mass, where it becomes increasingly hotter as it gets closer to the thermal source below, and more acidic as it interacts with the sediments and basalt. Eventually, once it reaches a critical temperature (400–500 C), it becomes buoyant, and moves back towards the surface. Research has indicated that possible seawater interaction with volcanic rocks occurs down to a depth of several kilometres below the seafloor. At this stage, the hot fluids have scavenged many elements from the oceanic crust, and are relatively enriched in economically-significant phases such as Cu, Zn, Ag and even Au, as well as typical rock-forming elements such as Fe, Mg, Si and the eponymous S. As the water reaches the surface once again, the rapid change in temperature and chemical environment causes the dissolved elements to crystallise under and on the sea floor, forming a series of mounds and hydrothermal vents. If this so-called hydrothermal fluid collected enough economically-important elements, and the chemical and thermal state of the sea floor is perfectly suitable for sufficient precipitation, then these elements may form significant accumulations of minerals (such as chalcopyrite with Cu and sphalerite with Zn) which can subsequently be located and utilised as an economic reserve. It is the exact combination of these physical features and chemical processes, occurring in the correct order for a long enough time, that allows the formation of sea floor massive sulphides (SMS); the chances of this happening are very low. These deposits may be surrounded by a halo of iron-rich sediments, and a plume of iron and manganese-rich particulates can develop in the overlying seawater which can extend and precipitate oxides many kilometres away from the hydrothermal source. See e.g. Robb (2004) and Chapter 3 for further details.

Six licenses have so far been issued by the International Seabed Authority for the exploration of sea floor massive sulphides in international waters; two along the slow-spreading (full spreading rate of between 20 and 55 mm/year) ridge further south in the Atlantic Ocean and four in the Indian Ocean (ISA, 2018). See Figure 1–4 and Figure 1–5.

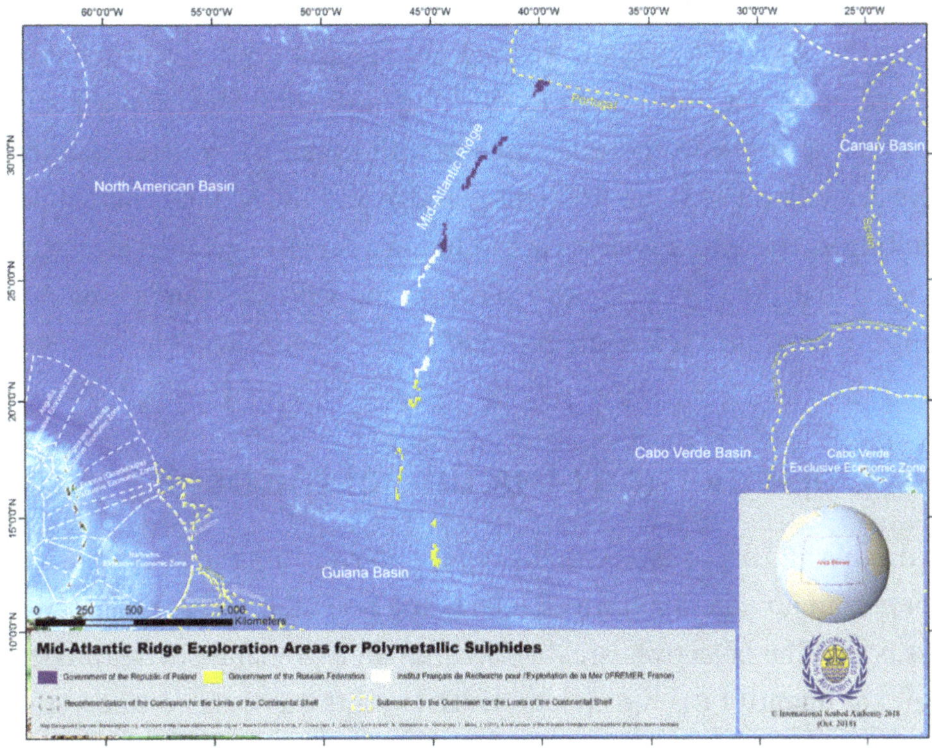

Figure 1-4 Russian, Polish and French exploration licenses issued in the Atlantic by ISA (Image: © International Seabed Authority). Click here for larger image.

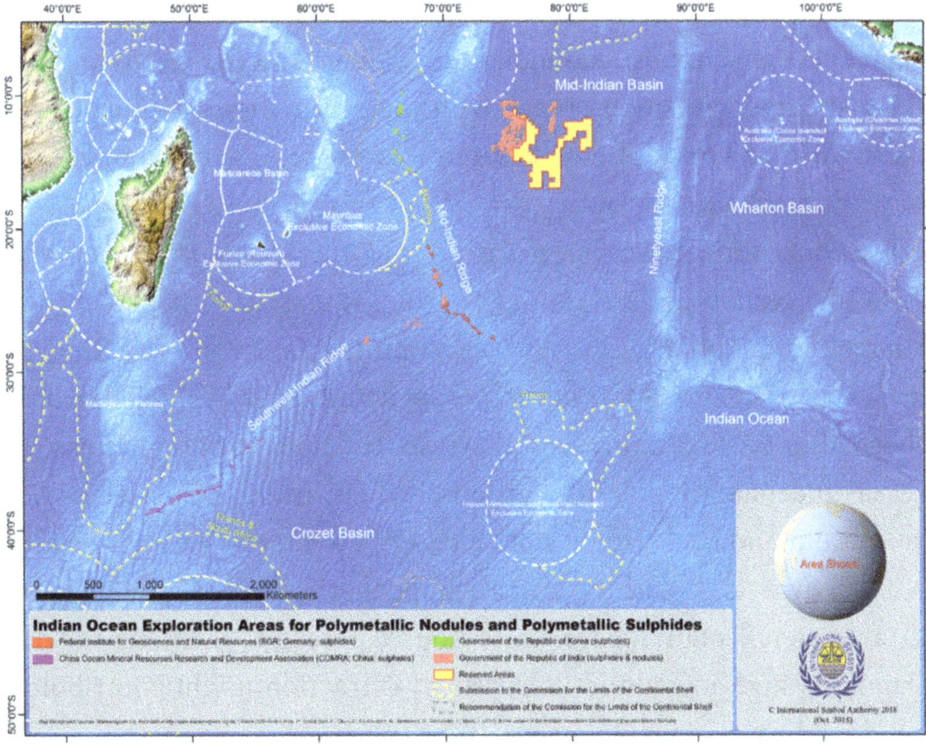

Figure 1-5 Exploration licenses issued in the Indian Ocean (Image: © International Seabed Authority). Click here for larger image.

Several active and inactive hydrothermal systems have been discovered along the AMOR (Pedersen et al. 2010a, 2010b). These vent fields are located inside the Norwegian Extended Continental Shelf. Two inactive and four active hydrothermal systems with potential mineral resources have been detected at the Kolbeinsey Ridge and the Mohns and Knipovich Ridges. These vent systems are found in a range of geological settings, ranging in depth from 140–2500 m depth, and with vent fluids ranging in temperature from a few degrees to 320°C.

1.4 Objective and structure of the book

Hydrothermal deposits , a specific type of marine mineral, located within Norwegian jurisdiction have an interesting potential and are the focus of this book. The objectives are, by focusing on a probabilistic play assessment, to quantify the resource potential and the associated uncertainty along the mid-ocean ridge inside Norwegian jurisdiction and to provide insights useful in future prospecting and exploration activities. We cannot, and do not, claim that all the inputs into the analysis are correct in the strictest sense, but with the probabilistic input we cover a realistic sample space (minimum and maximum values) and the resource potential is compared with other potential estimates and thereby validated. The majority of the work presented herein was executed during a period of two years, from 2012 to 2014.

This book aims at an international and interdisciplinary audience. However, a certain understanding of geology, marine minerals, mineral resource estimation and statistics is recommended in order to draw full benefits from the text. Some key terms are defined and explained in boxes imbedded in the text in order to increase readability and to increase the level of understanding. With this audience in mind, some terms have been left out and some might not be described in great enough detail to satisfy all experts. If nothing else, we hope that anyone reading the book will, at least, reflect on the mineral resource potential on and in the deep ocean floor, and how it and its potential extraction might contribute to meet the mineral and metal needs of the future.

The book initially gives a short geological background with focus on sea floor massive mineral resources. This is followed by an explanation of the play analysis methodology. One of the main challenges in play analysis, as it is used here, is to define the so-called permissive areas or tracts. How these are defined and their characteristics are explained in Chapters 4 and 5. Assessment results are presented in Chapter 6 and validated in Chapter 7. Chapter 8 describes the role of projects and activities like the MarMine Project (a research project funded by the Norwegian Research Council to study and sample one of the known active SMS systems) in validating the potential estimation results obtained in this and future studies. This chapter also includes a reference to the publicly-available results from two exploration cruises conducted by the Norwegian Petroleum Directorate, in 2018 and 2019. Some concluding remarks regarding resource potential are made in Chapter 9, along with a comparison and discussion of the results presented herein and the results from similar studies performed post 2014.

The book is the outcome of a collaborative project with participants from the Norwegian University of Science and Technology (NTNU), Nordic Mining ASA and Statoil ASA (now Equinor ASA) carried out between 2012 and 2019, and funded by Nordic Mining ASA and Statoil ASA. Steinar Løve Ellefmo and Fredrik Søreide have written most of the text and edited those portions drafted by the other researchers involved: Cyril Juliani, Krishna Panthi, Richard Sinding-Larsen and Ben Snook, all from NTNU, and Georgy Cherkashov, Sergey Petukhov and Irina Poroshina from VNIIOkeangeologia (Russian Institute for Geology and Mineral Resources of the Ocean).

Cyril Juliani and Ben Snook contributed with descriptions of the geology along the mid-Atlantic ridge, the geological processes and the analysis and characterization of sample material collected during the MarMine-cruise.

Richard Sinding-Larsen, along with Steinar Ellefmo and Fredrik Søreide, carried out the fundamental work in converting the analysis of the bathymetric data performed by researchers from VNIIOkeangeologia into the resource estimate. The results from this work were delivered to the project as a report referred to throughout the book as 'Ellefmo & Søreide, 2014'.

Krishna Kanta Panthi contributed the section 'Introduction to rock mechanics', which connects geodynamic analysis with rock mechanic principles, and he rewrote the 'Geodynamic analysis' section in chapter 4, 'Definition and description of permissive tracts in the absence of data'. He also contributed to the section on 'Morphostructural analysis'.

Georgy Cherkashov, Sergey Petukhov and Irina Poroshina carried out the analysis of the bathymetric /morphostructural data and thereby did the fundamental work necessary to define the permissive/favorable areas that were the input into the resource estimate. The results from their work were delivered to the project as a report referred to throughout the book as 'Cherkashov et al., 2013'.

Geological Background

2.1 Regional setting and ridge morphology

2.1.1 Introduction

The Earth is dynamic. Landmasses and ocean floors have been in motion since the dawn of time and will continue to be into the (un) foreseeable future.

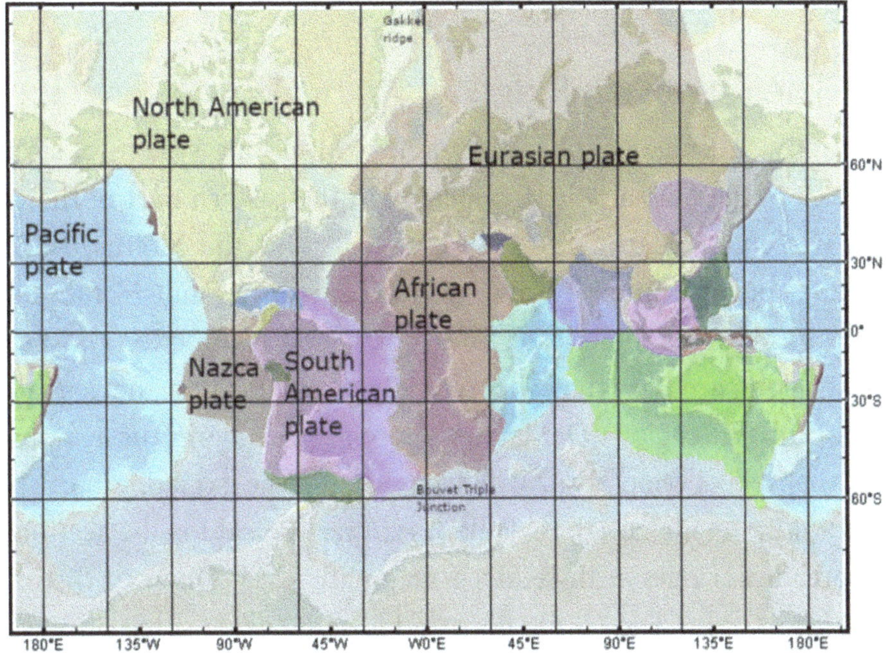

Figure 2-1 World map showing the landmasses, plate boundaries and mid-ocean ridges that run through the Atlantic, the Indian and the Pacific Oceans. Source: GeomapApp (http://www.geomapapp.org) version 3.6.8 and Global Multi-Resolution Topography Data Synthesis (GMRT Image Version 3.5). Click here for larger image.

Continents and ocean floors are destroyed at destructive plate boundaries where two plates meet, and new crust is made at divergent or constructive boundaries. We find destructive plate boundaries, for example, off the coast of Japan, where the Pacific Plate collides with and is pushed under the Eurasian Plate, and off the coast of South America, where the Nazca

Plate is pushed under the South American Plate and forms the Andes Mountains. The mid-ocean ridges belong to the longest mountain range in the world, constituting worldwide 64,000 kilometre long, volcanically-active and constructive plate boundaries. Water depths average about 2500 metres to the top of the ridge, reaching their deepest point of about 5500 metres in the Cayman Trough in the Western Caribbean and reaching above sea level in Iceland. In Iceland, the Mid-Atlantic Ridge (MAR) that splits the Atlantic Ocean floor into two parts, from the Gakkel Ridge northeast of Greenland in the north to the Bouvet Triple Junction in the south, crosses the Þingvellir National Park. The MAR separates the Eurasian and North American Plates in the north and the African and South American Plates in the south. See Figure 2–1.

2.1.2 The Knipovich Ridge

The Knipovich Ridge is an independent 550 km long system of spreading ridges in the North Atlantic, located in the eastern part of the Norwegian-Greenland Basin. See Figure 2–2. It joins the Mohns Ridge in the south, and is limited by the Molloy Transform Fault in the north. Its geological strike, practically in a north–south direction, is skewed (35°–65°) compared to that theoretically predicted for the spreading direction in this region (DeMets et al., 1990). All along the Knipovich Ridge, the well-developed rift valley has an average axial depth of about 3.5 km (Okino et al., 2002); its valley is rather typical for the sections of slow-spreading ridges with reduced magmatic feed. One can distinctly identify the 7–15 km wide inner floor and usually two or three ridge walls, often controlled by detachment faults, separated by rather wide terraces.

> A **detachment fault** is a large-scale fault structure associated with extensional tectonism. They can show tens of kilometres of displacements and play an important role in the formation of hydrothermal activity.

The ridge flanks that form the meridionally-extended plateau consist of volcanos and large massifs of complex origin, which are possibly oceanic detachments (Zonenshain et al., 1989). The eastern flank of the Knipovich Ridge is more subsided compared to the western side, a characteristic usually attributed to the accumulated sedimentary load (Kandilarov et al., 2010). On the other hand, no strong asymmetry is seen in the flange structure apart from the area north of 78 degrees north, where the eastern portion of the ridge valley joins directly with the continental margin of Svalbard. These characteristics of the Knipovich Ridge add a significant contrast to most other mid-ocean ridges situated far off continental margins. The system is situated close to the Svalbard and Barents Continental Margins, which were lifted up in the Late Cenozoic, some 60 million years ago. The area connecting the Mohns and Knipovich Ridges is close to the Bear Island Fan, which has a maximum thickness of more than 3 km of quaternary glacial sediments (Fiedler & Faleide, 1996; Faleide et al., 1996). The distal part of the fan reaches the spreading ridge. At the Mohns-Knipovich Ridge Bend, six multichannel seismic profiles have been collected and analysed in 2001 (Bruvoll et al., 2009). To the north of the bend, the Knipovich Ridge is situated close to the Bellsund and Isfjorden Fans, which have been in front of troughs on the continental shelf. About five million years old (late Plio-Pleistocene) fan complexes (Vorren et al., 1998) input sediments into the rift valley (Amundsen et al., 2011).

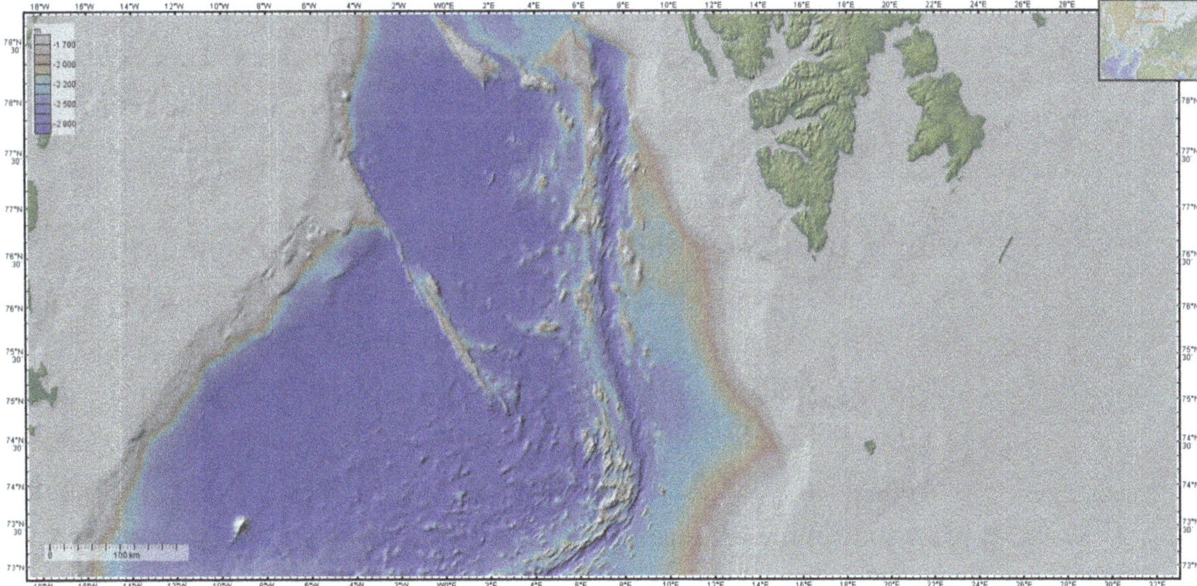

Figure 2-2 Knipovich Ridge characterised by a deep axial floor, high relief, highly oblique spreading and development of axial volcanic ridges. See inset for location. Source: GeoMapApp / GMRT Grid (Ryan et al., 2009). Click here for larger image.

Along the ridge, individual spreading segments are several tens of kilometres long and consist of two types: segments sustained by magmatic inputs and spaced at intervals of 85–100 km (Okino et al., 2002), and segments with lower relief that lack magmatism. Robust magmatic segments are considered to be loci of recent volcanic activity supported by mantle upwelling cells. The larger segment spacing observed on the Knipovich Ridge may reflect the large-scale thermal structure of the mantle that controls the focused magmatism. However, shallower geometric and kinematic factors, such as rift obliquity and plate separation rate, may also influence these processes. In particular, the southern Knipovich Ridge has lower relief spreading segment centres associated with high obliquity, slow spreading rates and reduced mantle melting, which results in a lower magma supply (Okino et al., 2002).

For the majority of the valley sections, one can hardly use the term 'volcanic segment' in its classical meaning, even though this issue is treated differently by different scientists, such as Crane et al. (2001) and Okino et al. (2002). It can be supposed that the morphostructural expressions for the Knipovich ridge develop in two stages, which is elaborated in the following. The topographic signature of breaks between segments, known as non-transform discontinuities, are clearly manifested on the valley flanks, along the zones running north-west and along the trend of the Paleo-Spitsbergen zone of displacement off-sets, which existed before the supposed beginning of spreading in this region (Crane et al., 2001). These signatures are expressed either by steep scarps with strike orientation having opposite inclination at the counter slopes of the valley or by sharp wedging of the axial structures directed north-east. In the first case, two systems of oppositely-inclining scarps intersect in the zones of non-transform displacement at the angle of 50° to 60°. Under these conditions, the elevation amplitude and volcanic activity are much higher. Thus, the tectonic processes at the segment boundaries control the arrangement of the areas of maximal magmatic activity, which clearly tends towards the zones of increased crustal permeability.

A **geomorphological (morphostructural) analysis** is an analysis of the morphology of the ocean floor that enables the identification of structures like flat-topped volcanoes, scarps, faults and axial volcanic ridges (AVRs), indicative of the probable presence of hydrothermal manifestations.

The Knipovich Ridge drops out of the formation scheme of an ordinary mid-ocean ridge. It lacks linearly-ordered magnetic anomalies (see Figure 2–4) and the major morphostructural elements are rather vaguely expressed. All this does not exclude a possible formation of sulphide ores on the Knipovich Ridge. Tectonism, as well as magmatism, are actively developed. Hence, hydrothermal venting followed by sulphide mineralisation is probable. Special studies are, however, necessary to reveal the criteria for predicting sites of possible SMS accumulation. These challenges are in this book addressed using geomorphological (morphostructural) analysis of the relief of the Knipovich Ridge and geodynamic modelling.

Basaltic lava flows of two age generations are visible on side-scan sonar images (Crane et al., 2001). Side-scan sonar is a method whereby acoustic images are acquired of the seafloor, often in great details, and show where young volcanic features and lavas are, or where the seafloor is covered in sediment. The youngest flows, scarcely covered with sediment, are less than several thousand years old (the sedimentation rate is 1 cm per 1000 years). In the northern part of the ridge, the young, sheet and pillow lava basalts are nearly equal and cover about the half of the valley floor. The largest volcanic structures form several linear ridges, 25–30 km long, with heights between 500 and 800 metres. These ridges cross the valley floor at a small angle and merge with the opposite scarps forming the valley walls. Moreover, many small irregular volcanic ridges have been recorded, including vast fields of predominantly developed sheet lavas, narrow basalt strips associated with small fracture outflows, and central type volcanoes. In the majority of cases, the volcanic manifestations are associated with linear tectonic structures (small faults and fractures) on the valley floor, and rarely with the large marginal faults. This is typical in spreading areas, where tangential stresses are close to zero. The maximal recent volcanism is recorded in the two segments south of 77 degrees north. Two regions, with somewhat more restricted development of the young volcanic rocks, are located in the intervals 77°15'–77°35' and 77°45'–77°58 north. There are also several volcanic centres located south of 76° N (Tamaki & Cherkashev, 2000).

The recent tectonic activity on the inner floor of the rift valleys is typically found at the areas of young volcanism and at the areas with thick sediment covers. It exhibits itself in the formation of an *en echelon* (oblique) system of small (the first tens of metres) fault scarps interfaced with linear fractures. The general strike of these structures varies between 25° and 35°, which approximately corresponds to the regional stress field developed due to sea floor spreading. Their end sections are often bent in the north-south direction, which suggests the existence of overall shear stresses. On the slopes of the rift valley, the modern tectonic activity is concentrated mostly at the zones of intersection of the longitudinal and transversal dislocations, corresponding to the non-transform discontinuities forming the spreading segment boundaries.

2.1.3 The Mohns Ridge

The Mohns Ridge, by the features of its structure, morphology and development, may be regarded as a mid-oceanic structure formed in the regime very close to a common MOR. See Figure 2–3.

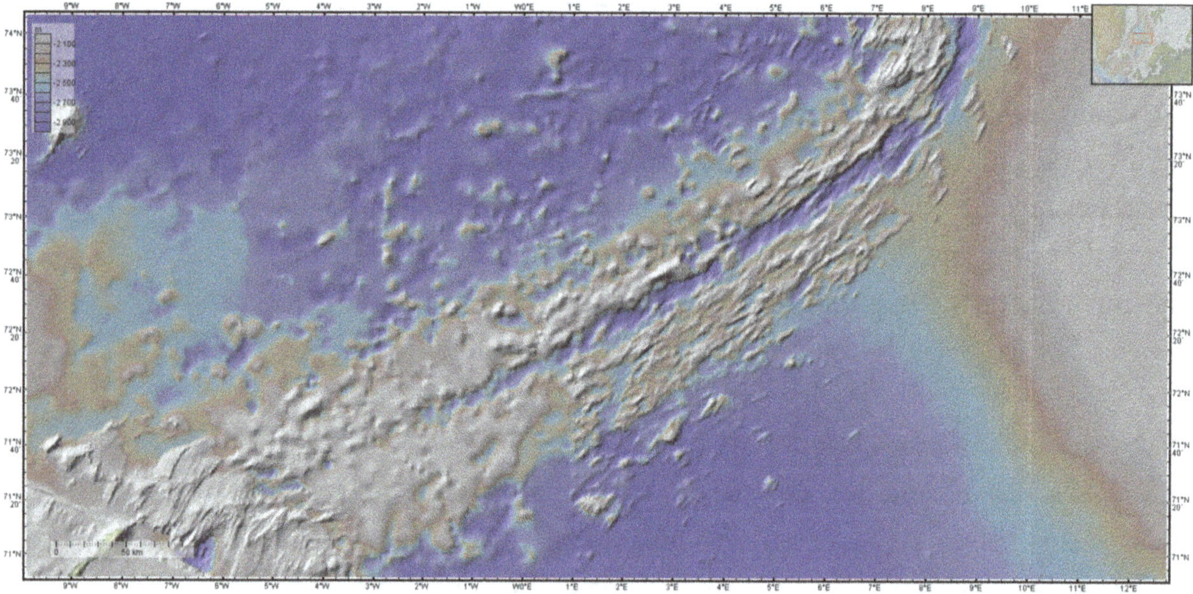

Figure 2-3 Mohns Ridge characterised by a deep axial floor, high relief, asymmetric development, oblique spreading and development of axial volcanic ridges. See inset for location. Source: GeoMapApp / GMRT Grid (Ryan et al., 2009). Click here for larger image.

The system of clear, linear magnetic anomalies indicates its spreading nature. See Figure 2–4. The absence of transform displacements and the skewed spreading are its abnormal features. Morphologically, it is well expressed. One may expect SMS type deposits in the valley among neo-volcanic elevations with rift structures, in the apical part on central-type volcanoes, at the edges with rotated blocks and flattened sub-horizontal terrace-like steps, outside the edges, between their upper rim and the toes of rift mounts. Several venting sites have been discovered in the region (Pedersen et al., 2010b). See Chapters 3.4 and 3.5.

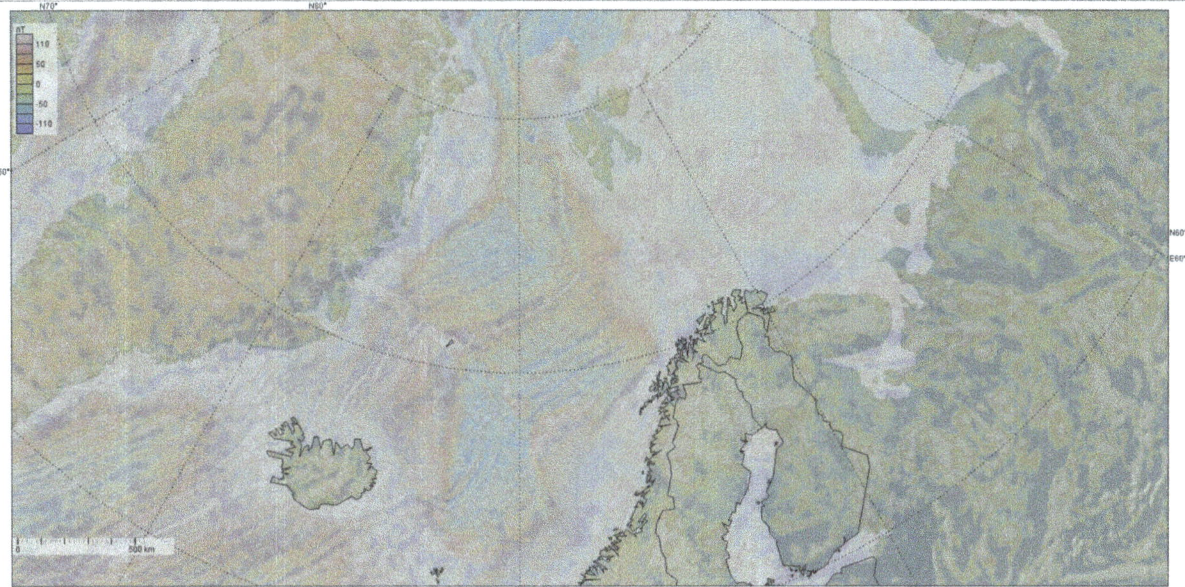

Figure 2-4 The magnetic anomalies on both sides of Mohns Ridge indicated by the blue and green repetitions indicating a varying direction of the magnetic field are not seen around Knipovich Ridge. Source: GeoMapApp (http://www.geomapapp.org) version 3.6.8 and the embedded 2-arc minute EMAG2-data from 2013 (http://www.geomag.us/models/emag2.html). Click here for larger image.

Mohns Ridge is an ultra-slow-spreading centre (<20 mm a^{-1}) situated within the Norwegian-Greenland Sea, that started opening approximately 53–55 Ma ago during the extension of the Mid-Atlantic Ridge (MAR) north of Iceland (Talwani & Eldholm, 1977; Torsvik et al., 2001; Vogt, 1986). It extends WSW-ENE for about 550 km from its southernmost part, delimited by the Jan Mayen Fracture Zone, to its northern junction with the Knipovich Ridge. The ridge presently spreads at a rate of about 15 to 16 mm/yr (Mosar et al., 2012) with an oblique relative displacement of 30° (Dauteuil and Brun, 1993). Its mean crustal thickness is of about 4.0 ± 0.5 km, a thickness considered lower than average for normal crust formed along the MAR (Klingelhöfer, 2000), but typical for ultra-slow-spreading ridges that generally show thinner and variable crustal thicknesses (Chen, 1992).

Along the ridge, water depth of the value floor ranges from 2.5 to 3.5 km below seawater level (Géli et al., 1994), and decreases progressively towards the south-west due to its proximity to the Icelandic hotspot (Talwani & Eldholm, 1977; Vogt, 1986). A characteristic aspect of the rift bathymetry is the strong asymmetric configuration of the flanks delimiting globally higher summits on the north-western sides (Vogt et al., 1982, Johansen et al., 2019). Moreover, the axial valley and the bounding flanks are marked by a series of en-echelon ridges interconnected by non-transform offsets (NTOs, see Figure 2–3) that commonly result from varying obliquity (Géli et al., 1994). This segmentation pattern occurs with intermittent volcanism characterised by basement highs elongated at the scale of approximately 20–30 km and oriented perpendicularly to the spreading direction. Previous studies distinguished these ridge segments to be volcanic edifices (Géli et al., 1994), while some involved tectonic origins solely with transtensional bounding faults (Dauteuil & Brun, 1993, 1996).

> **Non-transform offsets (NTOs)** are areas with a gradual transition between spreading segments without any significant transform faulting. They are discontinuities (second-order) dividing the spreading centre into segments at intervals of 10 to 100 km. They represent significant descriptors of the tectonic processes along the MAR (Grindlay et al. 1991 and 1992; Sempéré et al. 1993; Gràcia et al. 2000).

Investigation of the bathymetry shows that the width of the axial rift valley, i.e. the distance between the rift walls, ranges in average from 5 to 15 km. The axial valley often deviates at NTOs close to deep basins (≥ 3000-m depth). These basins usually lack recognisable structures or volcanic edifices, but are commonly bounded by large, dextral east-west trending transverse faults possibly accommodating the depth variation of the brittle/ductile boundary (Dauteuil & Brun, 1996). NTOs have limited lateral structural expression, but their off-axis trace remains visible from alternating sets of continuous ridge-parallel terrains and from structures that trend slightly obliquely to the spreading direction. The volcanic activity at these transitions appears to be very restricted, and the importance of such zones for hydrothermal venting remains to be shown.

In the south-western part of the ridge, segments and transitions are difficult to identify because of irregular and complex bathymetric patterns associated with abnormally abundant volcanic activity, as well as to curved tectonic structures. The melting anomaly associates with more-shallow and smooth terrains as a result of the southward, hotspot interaction (Talwani & Eldholm, 1977; Vogt et al., 1981), whereas the curved structures respond to oblique extensions. In contrast, the northernmost Mohns-Knipovich Ridge Bend consists of a more symmetric ridge segments expressing a rough topography at the higher western side compared to the sediment-covered eastern flank, where only a few basement ridges are exposed. Along the overall ridge, the western flanks are, on average, 200 m higher than the opposite side, due to higher tectonic activity; their edges may sometimes correspond to rotated and uplifted fault blocks forming downward curved ridge flanks that are usually attributed to low-angle detachment surfaces (Smith et al., 2008).

Apart from the anomalous segment adjacent to Jan Mayen, most of Mohns Ridge has a well-developed rift valley with average depths of 2000–3200 m. In the north-east direction, a general drawdown of the inner rift-valley floor and the flange highs are noticed. The western flank is about 500 m higher than the eastern one. Hence, at the western flank, the tectonic deformations should be much higher compared to the eastern one.

> **Grabens** are suppressions in the terrain or the ocean floor, caused by (normal) faulting in an extensional tectonic environment where the central part of the suppression sinks in compared to the flanks.

The morphology of the axial volcanic rises varies considerably, which may clearly be expressed by 400–800 m high axial rises. These rises are sometimes strongly deformed by grabens. There is an unusual abundance of large, flat-topped central-type volcanoes associated with different morphostructural settings in the valley floor and at its eastern flank.

The strike of the valley reaches 60 degrees, though the majority of its segments are separated by the right-slip offset and oriented at 40–50 degrees. The amplitude of its lateral offsets reaches 5 to 7 km in the northern part of the ridge, 8 to 15 km in the central part, and is rather irregular at the south where individual segments are often difficult to identify. In total, 15 spreading segments, with an average length of about 30 km, are recognised along Mohns Ridge.

2.2 Formation of seabed massive sulphide (SMS) deposits

Globally, there are some 64,000 km of mid-ocean ridges (MORs) where SMS deposits occur and, most of the time, these are outside the jurisdiction of national exclusive economic zones (EEZs). However, within the Norwegian extended continental shelf, the SMS deposits are formed at mid-ocean on the Mohns and Knipovitch Ridges, where Norway regulates exploration and potential mining of SMS deposits. Because of the extensional nature of MORs, melting due to pressure decreases produces mafic magmas that form new oceanic crust. The process involves crack propagation, where cold seawater flows in and hot, mineral-rich seawater flows out, creating chimney-like structures and plumes known as black smokers. Some of these cracks may extend thousands of metres into the lower crust. See Figure 2–5.

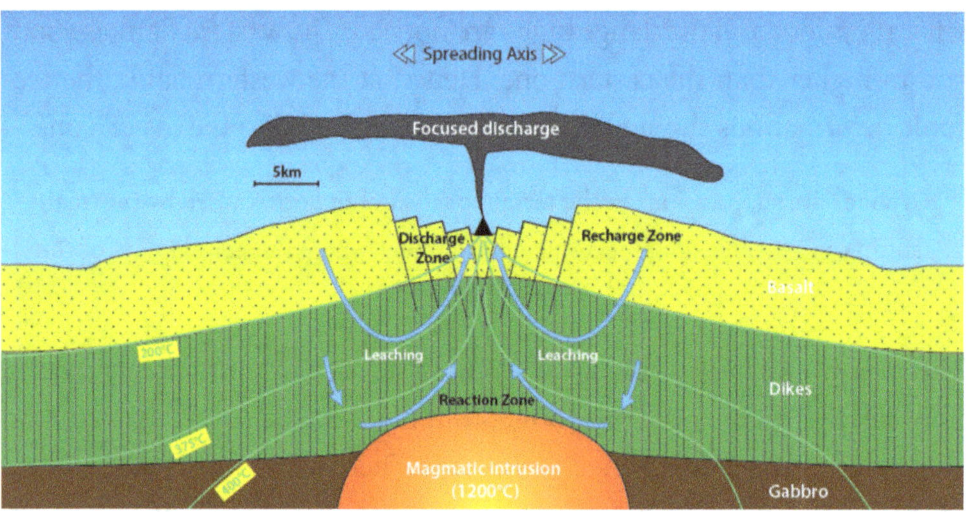

Figure 2–5 Formation of seabed massive sulphide deposits. A diagram depicting a mid-ocean ridge and hydrothermal circulation. Click here for larger image.

Conventional models of hydrothermal systems comprise sub-sea floor seawater circulations driven by a heat source, where seawater migrates downward through cracks and reacts with crustal or upper mantle rocks at several kilometres depth. During this process, seawater can be heated to temperatures of 400–500C, high enough to modify its physical properties and will eventually find its way upwards through features facilitating hydrothermal flow. This downward-upward convection scheme involves leaching of metals from the surrounding rock via chemical reactions. Correspondingly, the heated seawater evolves to a fluid more acidic, reduced and enriched in dissolved metals and sulphur. When these hydrothermal fluids rise to the sea floor, it expels most of its metal content in the water column through chimney-like structures (black smokers). A part of these components remains in the upper part of the sub-sea floor where metals precipitate into metal sulphides from conductive cooling and mixing of hot hydrothermal fluids with cold seawater. From these processes, metals of economic interest accumulate on the sea floor and underneath by mineralisation of the rock basement, which results in forming SMS deposits that contain significant amounts of sulphide minerals (e.g. Robb, 2004). These deposits may be comparable to or larger than some deposits on land (e.g. Scott, 2004).

SMS occurrences in the deep sea may occur at different tectonic settings like the mid-ocean ridges (MORs). Presently, all the known vents on slow or ultra-slow spreading centres, like the Mohns and Knipovitch Ridges, are distributed in various geologic settings; both magmatic and tectonic-dominated settings. The former are directly controlled by axial volcanic structures controlled by the presence of a high-temperature magma under the seabed. The latter are usually located at the more-or-less remote flankes of the rift valley, mainly in the areas of the development of large offset detachment faults forming structures called core complexes. In this setting, deep penetrating faults represent the dominant geological controlling process on hydrothermal circulations, and can, under certain circumstances, associate with underlying magmatic processes.

2.3 Factors controlling the location and the size of SMS deposits

Both volcanic and non-volcanic tectonic activity is usually long-lived at slow to ultra-slow mid-ocean ridges. This provides stability for hydrothermal systems (Hannington et al., 2005; Rona, 2008). The estimated frequency of venting sites along the mid-ocean ridges varies from 1 to every 400 km to 3 per 100 km (e.g. Baker & German, 2004; Hannington et al., 2010, 2011b). Faults play a vital role, both in charging and discharging hydrothermal systems, by acting as fluid conduits. Hydrothermal manifestations may, therefore, occur on structural highs many kilometres off-axis. The Snake Pit hydrothermal field occurs at a neo-volcanic ridge (Fouquet et al., 1993), and the TAG field is situated approximately 3 kilometres from the axis (Rona et al., 1993; Kleinrock et al., 1996). Hydrothermal systems can be driven by both serpentinisation and magmatic processes. These could act in combination, as in the Rainbow (Fouquet et al., 1997) and Logatchev fields (Mozgova et al., 2005). Slow-spreading ridges are structurally complex, showing evidence of tectonic extensions, large amounts of faulting (both detachment faulting and normal faulting) and exposure of gabbros and partly serpentinised ultramafic rocks in oceanic core complexes (Smith et al., 2008).

2.4 Loki's Castle and Mohns' Treasure venting sites

Loki's Castle is a hydrothermal vent field at the ultra-slow-spreading Arctic Mid-Ocean Ridge (Baumberger et al., 2010; Pedersen et al., 2010a, 2010b). The hydrothermal fluids show indications of having leached both sediments and mafic rocks in the ocean crust for metals. See below and Baumberger et al. (2010).

The ore-bearing vent, named Loki's Castle, is located at 73°30' N., 8° E. It lies on an axial volcanic ridge formed by young pillow lavas. The chimneys are up to 11 m high and show a maximum temperature of the fluids of 317° C. The content of CH_4 and H_2 is interpreted to indicate the presence of ultramafic rocks at depth (Pedersen et al., 2010b). The geochemical properties are similar to those of the Middle Valley and the Endeavour Ridge in the Pacific Ocean.

Loki's Castle is partly overlapped by a sediment cover. The region features high tectonic stress – a combination of fault disturbance and domical rises providing potential channels for the fluids at the formation of Loki's Castle.

Samples collected show a maximum metal content for Fe at 31wt%, for Zn at 5.4wt% and for Cu >1wt%, and the sediments contain non-insignificant amounts of metals and average 4.6 wt% Fe, 100 ppm Zn and 33 ppm Cu (Pedersen et al., 2010b).

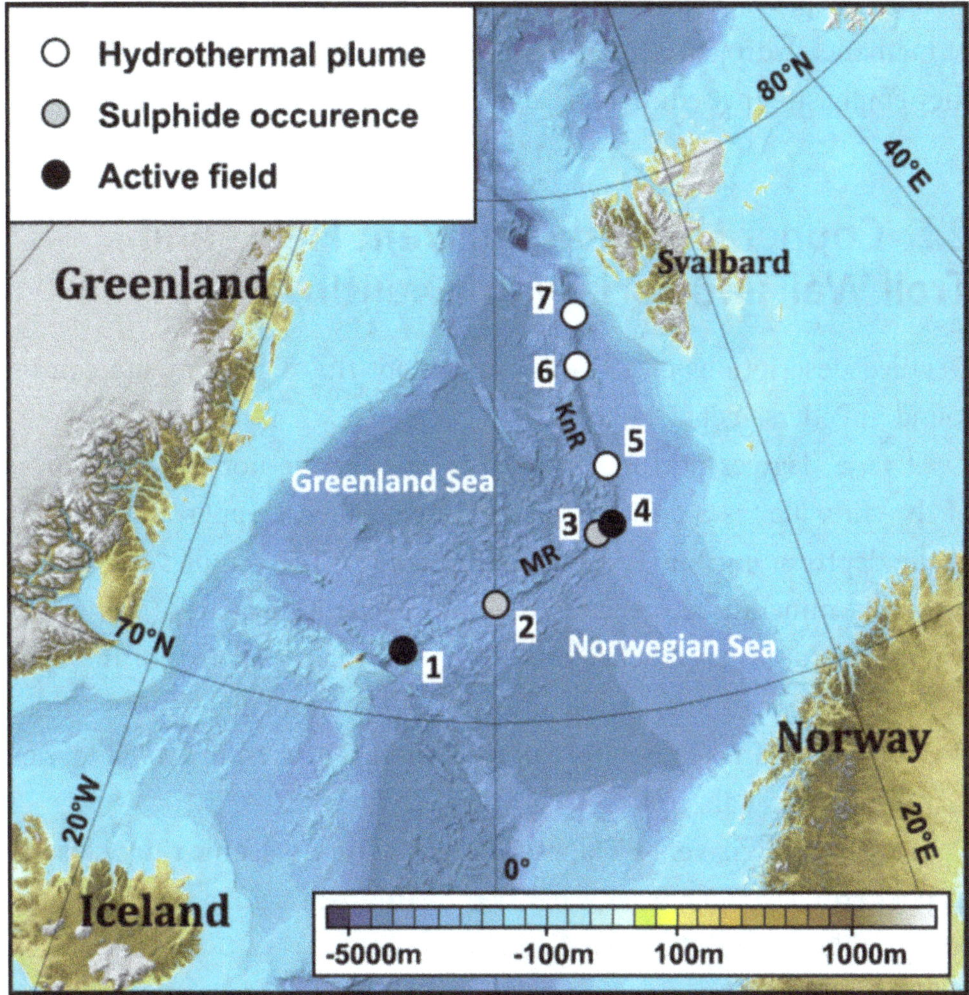

Figure 2-6 Active fields, sulphide occurrences and plumes along Mohns and Knipovich Ridges. MR = Mohns Ridge, KnR = Knipovich Ridge. Source: NTNU / Cyril Juliani. Click here for larger image.

1. Jan Mayen (Troll Wall, Soria Moria)
2. Copper Hill
3. Mohns' Treasure
4. Loki's Castle
5. 75°N. Plume with anomalies of LS, CH4, pH and ATP (Tamaki et al., 2001).
6. 76°48′N. Water-column temperature anomaly near north flank of submarine volcano (Sundvor, 1997).
7. 77°40′N. High CH4 and He-3 ratios observed. Potential occurrence of serpentinisation reactions (Connelly et al., 2002).

Baumberger et al. (2010) have investigated the chemical composition and the stable and radiogenic isotopes of the vent fluids exhaled from the Loki's Castle vents and quantified the associated signatures. These signatures are compared to both sediment-influenced and sediment-starved systems. A pH of approximately 5.5 and relatively high concentrations of methane, hydrogen and ammonia are reported. Based on their study, they emphasise the sedimentary contribution.

2.5 Copper Hill and Ægir Vent Fields and Troll Wall and Soria Moria venting sites

Detailed descriptions of all vent sites except the Ægir vent site can be found in Pedersen et al. (2010b).

At 72.32 degrees north, 2.10 degrees east, in the north-western part of the rift valley is a mineralised vent zone called Copper Hill, located at the depth of 900 m (Pedersen et al., 2010b). The samples are shown by highly-mineralised breccias. The main mineral is quartz (more than 50%). Sulphides (mainly chalcopyrite) make up around 30% of the total bulk volume. The minerals are altered. It is accompanied by iron oxides and malachite. The temperature of formation ranges from 270° C to 330° C (Cherkashov et al., 2013).

In 2015, the Ægir Vent Field was discovered on a volcanic ridge close to Copper Hill, at 2200 metres water depth. This vent field is reported to be comparable to Loki's Castle (Olsen et al., 2016).

Two vents at 71° north are located 5 km from each other in the utmost-south section of Mohns Ridge.. In terms of depth, it is the shallowest section and its most elevated points reach 500 m. Seismic surveys have shown that the crust is 10 km thick. The major volcanic structure associated with the Troll Wall is a volcano with a 5 km caldera built on glacio-clastites. A rift 150 m deep and 2 km wide transects the volcanic structure. The zone of venting is associated with the fracture of the eastern board of the rift. Deposits enriched in iron have been recorded here. The active part of the fracture is 1 km long. The vent consists of ten chimney structures 5–10 m high. The chimneys are distributed on the slopes of mounds of hydrothermal origin, and are emitting white smoke and CO2. The chimney material is composed of anhydrite, barite, sphalerite and pyrite. Around the focused sources, through the piles of blocks, diffused outflow of hydrothermal fluids can be seen. They are associated with bacterial mats, and the temperature of underlying sediments is 80°C (Cherkashov et al., 2013).

Soria Moria lies south of the Troll Wall, on the top of a volcanic edifice at a depth of 700 m below sea-level. The venting area is about 7 km². Two venting sites 100–200 metres in diameter each can be seen. The hydro-thermal deposits overlap lava flows. The smokers are of two types: white smokers are represented by multiple sulphurised chimneys 8–9 metres high. The others are transparent smokers of lower temperature. They are built of barite and silica with a small part of pyrite, sphalerite and galena. The chimneys are 10 m high and 15–20 m wide. The walls of the white smoker chimneys are covered with bacterial mats.

Both fields at 71° north feature high-temperature venting (270° C), a pH ranging from 4.1 to 5.2, and emissions of methane and CO2.

Play Analysis and 3-part Assessment

3.1 How undiscovered resources are estimated

The SMS resource assessment project on the Norwegian Continental Shelf estimates undiscovered resources with the aid of a method known as play analysis. This is a recognised approach, mostly used by the petroleum sector (NPD, 2013), but equally applicable to the mining sector (Juliani & Ellefmo, 2018b). Similar methodologies have been applied to assess mineral resource potential worldwide and in the USA for different commodities (Singer et al., 2010) as well as for oceanic seabed resources (Singer, 2014). The assessment process applied herein involves systematising and describing the SMS metallogenesis on the Mid-Atlantic Ridge, with additional information from appropriate ancient massive sulphide analogues. Play analysis shares many similarities with 3-part assessment (Singer et al., 2010). Five basic questions are addressed:

- Where and what represents the permissive tracts (favourable areas) for potential hydrothermal SMS resources?
  - The answer to this question includes a georeferenced region attributed an area in square kilometres.
- What are the chances that hydrothermal SMS resources exist inside the permissive tracts?
  - The answer to this question is a probability from 0 to 1 or 0 to 100%.
- If hydrothermal SMS resources exist and extensive exploration is performed, how many future accumulations will be found?
  - The answer to this question is basically a number. The number presupposes that the tract actually contains accumulations.

- What is the expected size distribution of future accumulations?
  - The answer to this question is the quantification of the min., max. and expected accumulation tonnage, including an assumption on the shape of the distribution.
- What types of metals and what grades will the future accumulations have?
  - The answer to this question is the naming of what elements and metals that are expected to be found and their grade distributions.

The SMS resources from both Mohns and Knipovitch Ridges are in this study regarded as coming from the same super-play. Each favourability area outlined in Chapter 4 will accordingly be modelled as a sub-play of the ridge super-play. Estimated resources are more uncertain the less is known about the SMS mineralisation process, and this project specifies accordingly resources with an uncertainty range. A probabilistic framework is chosen (GeoX, 2014) that can be used to estimate the amount of metals which could be proven and potentially produced from each sub-play given a deposit-modelling approach (Sinding-Larsen & Vokes, 1978). In this early assessment, the focus has been on aggregate contained metals rather than on grades and tonnages of individual SMS accumulations.

A **play** is a geographically and stratigraphically delimited area where a specific set of geological factors, such as recipient volumes, source rocks, leaching and hydrothermal flow paths, are present so that SMS deposits may be provable.

The project's estimate for undiscovered resources on the 1500 km MAR inside the extended continental shelf for Norway includes only the 1030 km of the ridge that has been interpreted by VNII Okeangeologia (Cherkashov et al., 2013). A small part of international sea floor intersects the assessment area on central Mohns Ridge, but without consequences for the assessment results in this study.

A permissive tract (Singer, 2010) is used in this work as a play.

Seventeen separate SMS sub-plays are defined along the ridges, based on morphostructural and geodynamical indicators. The plays encompass both the favourability areas that are outlined by the four morpho-favourable areas (MF-areas) on Knipovitch Ridge and the twelve morpho-favourable areas on Mohns Ridge, as well as the eight hydrothermal favourability areas (HF-areas) defined by discharged and stressed areas along the ridges (Cherkashov et al., 2013).

A play probability factor is a factor that needs to be present and effective in order for an accumulation to form. The two play probability factors (permissive tract probability factors) considered here are factors that combine the probability of having a metal and an energy source and effective migration of metal-laden hot fluids, and an effective precipitation and preservation of minerals and metals in or on the ocean floor:

1. The presence of an adequate source for metals, derived from the oceanic basement where a seawater hydrothermal convection system (migration) above a sub-axial magma chamber is capable of leaching elements such as Cu, Zn, together with Au, Ag from the crust.
2. The presence of adequate recipient volumes that could accommodate the precipitation and accumulation of sulphide concentrations, either on
   - neo-volcanic elevations within rift structures, associated with high-amplitude flank faults, e.g. axial volcanic ridges (AVRs)
   - in the central part of the central-type volcanoes
   - at the edges of the axial valley where angular relief blocks can be seen
   - on flattened sub-horizontal terrace-like steps, or
   - outside the spreading ridge flanks or between the upper rim and the toes of rift mountains with a large elevation contrast.

A confirmed SMS sub-play is characterised by the presence of at least one black smoker. Such a sub-play would have a play probability of 1, both for the adequacy of factor 1 as well as for factor 2. If the presence of black smokers has not been confirmed, the SMS sub-play remains unconfirmed. Since the sub-plays have been defined based on morphostructural elements referred to in adequacy factor 2, the confirmation of the SMS sub-play is associated with ensuring that the play probability for adequacy factor 1 is fulfilled.

The individual SMS probabilities at the local scale, given that the play is confirmed, are controlled by the following three principal conditions for the local presence of a potentially-producible massive sulphide accumulation within the SMS sub-play:

1. …that a potentially black smoker with a given minimum tonnage will be found if the sub-play is thoroughly explored, i.e. the existence of a local manifestation of hydrothermal activity
2. …that the hydrothermal activity indicates the presence of polymetallic sulphides (i.e. a black smoker)
3. …that the potential SMS accumulation at least has a tonnage and grade that corresponds to the defined minimum values.

3.2 SMS accumulation; deposit density and size (tonnage) and grades

The local existence of vent fields that may host a SMS accumulation is a fundamental component of a SMS play analysis. The number of vent fields and the amount of metals each of them could produce determines the estimated resources for each SMS sub-play. The estimation of the number of vent fields with hydrothermal manifestations in a sub-play therefore requires an unambiguous definition of what constitutes a 'SMS accumulation'. Unfortunately, as stated by Hannington et al. (2011a), descriptions of SMS accumulations in literature have included everything from a single vent or chimney to a large mound or cluster of vent complexes. It is, therefore, important to ensure consistency between the distribution of sizes and the density of accumulations (Singer, 2014). The well-mapped Troodos massive in Cyprus has, for this reason, been chosen as an analogue for the mineralisation on the slow-spreading Mohns and Knipovich Ridges. Compelling arguments that the Troodos ophiolite is a valid analogue for MAR have been advanced (Granot et al., 2011), where it has been demonstrated how the Troodos ophiolite can unveil the process of gabbroic intrusions along a slow-spreading ridge.

In this study, information from the Cyprus analogue (Singer & Mosier, 1986), from MAR deposits (e.g. Fouquet et al., 2010 and Cherkashev et al., 2013) and global grades for MOR (Hannington, 2013a, 2013b) have been used to model vent SMS field sizes and metal contents. All sub-plays have been modelled with the same tonnage and grade distribution, due to a lack of information that could be used to differentiate between the sub-plays.

Definition and Description of Permissive Tracts in the Absence of Data

4.1 Definition of permissive tracts

4.1.1 Introduction

This sub-chapter introduces rock mechanics and describes briefly how this is used in this study in geodynamic analyses, to highlight de-stressed zones prone to show hydrothermal manifestations. It continues with the basis for the morphostructural analysis and its link to deposit probabilities. Sub-chapters 4.1.3 and 4.1.4 describe the different permissive tracts that have been defined based on both geodynamical and morphostructural analysis. Details of the 17 sub-plays are provided in Sub-chapter 4.2. The identification of zones favourable for hydrothermal vent activity is based on the following indicators:

a. tensional structures (purple dotted line in Figure 4–12 and Figure 5–13) in the stress distribution maps associated with the highest gradient (black line in Figure 4–12 and Figure 4–13)

b. locations where the tangential stress acting on any fracture, fissure or discontinuity in the upper layer (see Sub-chapter 4.1.3) is close to zero (yellow line in Figure 4–12 and Figure 4–13. Area where a) and b) coincide is marked with a red line / red area in the figures.)

c. overlap between points described in a) and b) above and morphostructural hydrothermal indicators, like axial volcanic ridges with graben structures, structures associated to detachment faults and ocean core complexes, longitudinal faults, transversal dislocations and large flat-top volcanoes.

Delineation of the most prospective areas for finding SMS, based on the geodynamical analysis, are outlined as white boxes in Figure 4–12 and Figure 4–13 . Prospective areas outlined based on the morphostructural analysis are shown in Figure 4–16 and Figure 4–17 with yellow dotted lines. These yellow dotted lines have formed the basis for the definitions of the permissive tracts.

4.1.2 Introduction to rock mechanics

The term mechanics means the study of the equilibrium and motion of bodies, which includes static and dynamic forces. Modern rock mechanics deals mainly with engineering application of a rock mass to civil, mining, petroleum and environmental engineering (Harrison & Hudson, 2000). However, the Earth's dynamic body can also be related to rock mechanic principles, because there is a considerable influence on rock mechanic principles by natural processes that occur in the Earth's crust. This comes from plate tectonic movement, resulting in folding, faulting, spreading and mountain-building processes as indicated in Figure 4–1.

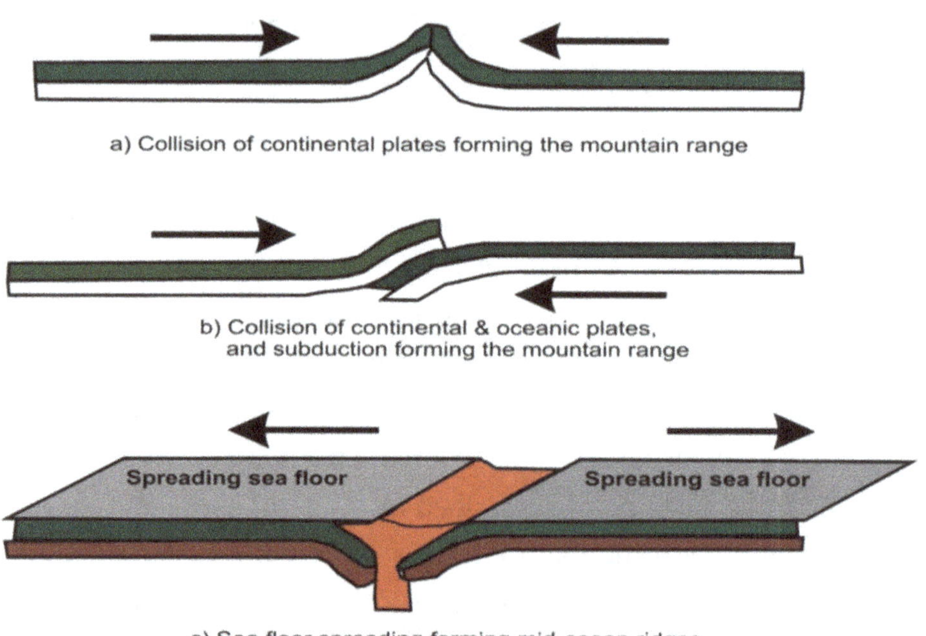

Figure 4–1 Conceptual mechanism of plate movement in the Earth's crust (Panthi, 2016). Click here for larger image.

Therefore, rock mechanics is a discipline that deals with forces that influence or change stress and strain environments in the rock mass. As seen in Figure 4–1, there are three main dynamic processes that are active in the Earth's crust:

a. mountain ranges are formed due to the collision of two continental plates of similar density, such as the Caledonides Mountain Range in Scandinavia
b. mountain ranges are formed due to the collision of continental and oceanic plates and subduction of the oceanic plate beneath the continental plate
c. formation of mid-ocean ridges due to sea floor spreading, a tensional environment where tangential stresses close to zero can be found

These dynamic processes influence the rock mass of the surrounding areas of collision, subduction and spreading considerably. These influences may be explained by the orientation of in-situ stresses in the vicinity of concern (Figure 4–2). These are mainly related to the dynamic tectonic movement of the Earth's plates. According to Panthi (2012), the magnitudes of total tectonic horizontal stress vary considerably and depend upon geographical location, geological environment (homogeneity in the rock mass), and distance from the main tectonic fault systems.

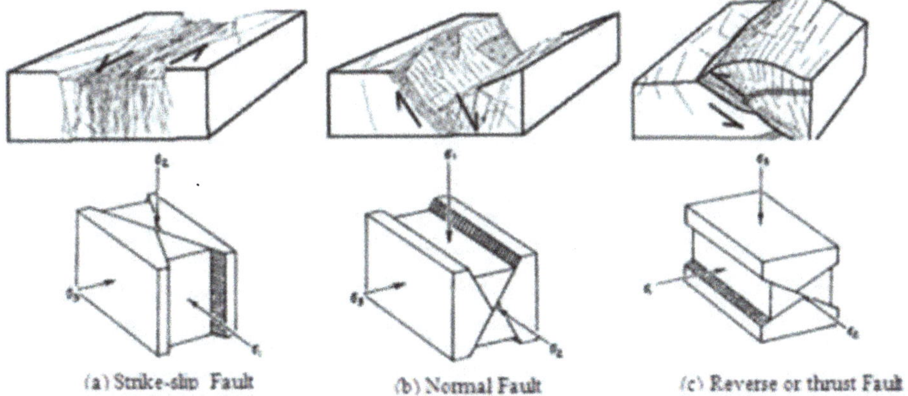

Figure 4-2 Anderson's classification of faults (in Panthi, 2006; modified from Braathen & Gabrielsen, 2000; and Rowland & Duebendorfer, 1994 . Illustration © K. K. Panthi; no reuse without permission). Click here for larger image.

45

In the course of the formation of mid-ocean ridges, many splays of long discontinuities (fracture systems) are formed, following the principle of the strike-slip faulting system (Figure 4–2a) caused by sea floor spreading, where major principle stress (σ_1) and minor principle stress (σ_3) are where the horizontal and the tangential stresses are close to zero. Figure 4–3 shows tectonic stress orientations worldwide.

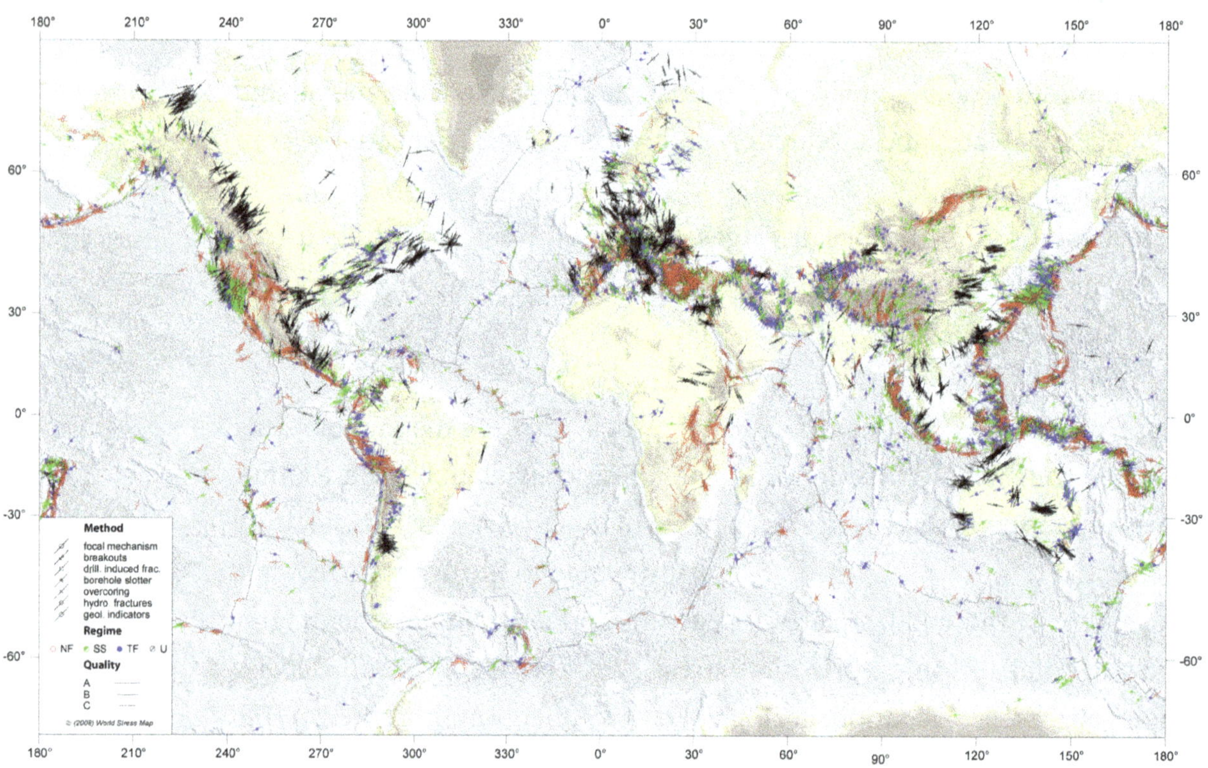

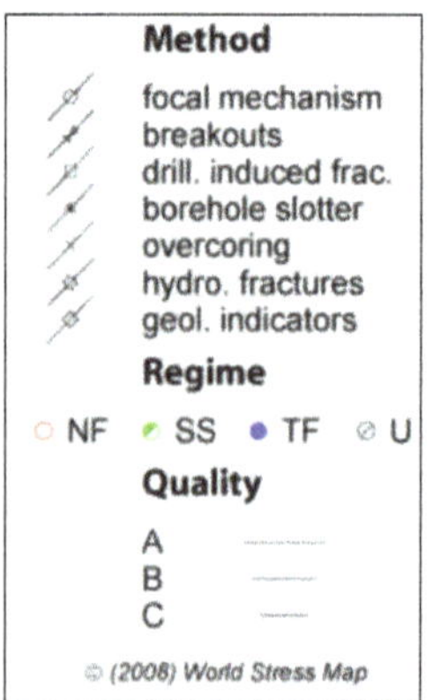

Figure 4-3 World Stress Map showing tectonic stress orientation and indicating plate division (Source: Heidbach, O., et al. (2018). The World Stress Map database release 2016: Crustal stress pattern across scales. *Tectonophysics, 744,* 484–498. http://doi.org/10.1016/j.tecto.2018.07.007). Click here for larger image.

This principle can be used to detect the areas that are prone to contain valuable minerals connected to sea floor massive sulphide deposits in the mid-oceanic ridge region. This chapter deals with assessing the possibilities for such anomalies along the Arctic Mid-Ocean Ridge (AMOR).

The tectonic stresses may be related to the current tectonic activities caused by the movement of tectonic plates. On one occasion, the rock mass may be strong enough to sustain these movements without any failure in the rock mass, leading to the storage of a considerable amount of in-situ tectonic stresses. On other occasions, the rock mass may become completely fractured, crumbled and folded, leading to a de-stressing situation. This is the reason why many regions in the world, where tectonic plates are in a state of limiting equilibrium due to strong rock mass, have high horizontal stress magnitudes, like in Scandinavia.

All three processes, shown in Figure 4–1, contribute to the overall stress state at a given point in a rock mass. In addition, other factors can alter the stress state, such as erosion, which will reduce the vertical stress component more than the horizontal components. Furthermore, in a condition where tangential stresses have magnitudes close to zero, a tensional environment will be created causing fractures at all scales and a typical de-stressed field.

Similarly, the concept of strain needs to be understood, so that rock mass deformation can be quantified. There are many applications of strain analysis. For example, some in-situ stress determination techniques measure strain and then compute the stresses. Therefore, it is necessary to be able to analyze the strain values so that the understanding of the anticipated deformation in rock masses can be understood. Strain is a measure of the deformation of a body and should be taken as being the same type of mathematical quantity as the stress. These strains, therefore, may be either normal strains or shear strains, which are directly analogous to normal stresses and shear stresses. Mathematically, they are linked through Hook's Law (Hudson, 2000).

4.1.3 Geodynamic analysis

Deformation modelling can be conducted to identify stressed areas of the crust where hydrothermal fluids may be discharged. The magnitude of stresses is calculated by geo-hydro-mechanical modelling. Such modelling helps to establish criteria correlating stressed zones of hydrothermal activity in the crust, which represent areas of tension, compression and discharge. The final aim of such modelling is to predict areas which potentially contain SMS deposits.

The mathematical framework follows Ludwig Prandtl's theoretical formulation of interface hydrodynamics (introduced by Prandtl, 1904), which states that the optimal channel configuration for fluid flow is circular and occurs at areas where tangential stress is zero (τ_t =0), see Figure 4–4.

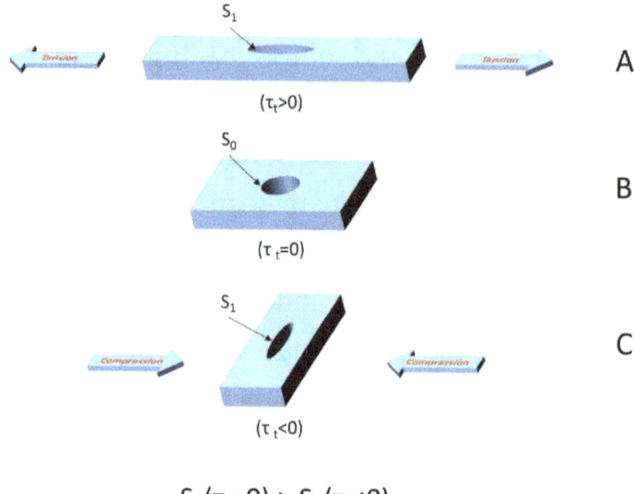

$$S_0(\tau_t=0) > S_1(\tau_t\neq0)$$

Figure 4-4 The influence of tangential stress τ_t on the cross section of a fluid flowing channel. Top: tension; Middle: Discharge; Bottom: Compression . Click here for larger image.

Thus, the search for the hydrothermal areas with the most vigorous discharge of multi-phase fluid involves defining areas of the sea floor (Figure 4–1) where tangential stresses in the upper layer are near zero (Figure 4–5). In the model, the real sea floor relief is approximated by a double-layer structure, consisting of an elastic (upper) and a plastically deformable (lower) layer. Under the weight of the upper layer, the lower layer suffers plastic deformation where stress magnitude is directly related to the accumulated overlaying mass (Figure 4–5).

Figure 4-5 Double-layer structure, consisting of an elastic (upper, 1) and a deformable (lower, 2) layer. Click here for larger image.

The mechanical properties, consisting of modulus of elasticity, Poisson's ratio, specific weight of the model, are estimated based on the properties of a 5 kilometer (Klingenhöfer et al., 2000) thick oceanic crust, composed of tholeiitic basalts similar to those that Iceland is built up of. These are considered typical for the areas of the Arctic Mid-Ocean Ridge (AMOR). The assumption made is that the properties of this 5 kilometer stretch of basalt represent the average properties of the AMOR crust. Considering the complex physical structure of the oceanic crust, with basalts, sheeted dykes, gabbros and ultramafic rocks, this represents a simplification.

The real depth of the elastic layer is dependent on the thermal state of the crust. Young and warm crust is thinner than old and cold crust (e.g. Klingenhöfer et al., 2000). This has not been taken into account in the model developed here.

The AMOR contains a major fault system and has considerable influence on the surrounding area, where numerous faults have been and are being developed. These faults represent de-stressed areas, where compressional horizontal stresses caused by far-field stress are close to zero. Hence, only the horizontal compressional stress caused by gravity alone is relevant in the analysis.

In the model, the elastic layer is represented as an elastic membrane deflected on rods, with lengths and positions defined by the coordinates (x, y and z) at the point of contact between the rod and the membrane (Figure 4–6). The length is given by the z-coordinate. The coordinates are taken from the bathymetric data with a grid size of 100 x 100metres.

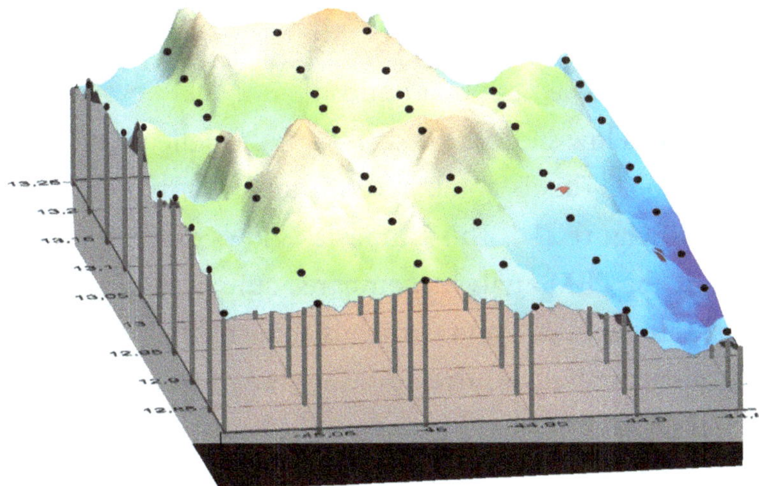

Figure 4-6 The simplified two-layered model of the Earth's crust used as a basis for the tangential-stress estimations. Click here for larger image.

The stress distribution in this model is calculated using the Finite Element Method (FEM). The idea of FEM is to model a continuum by discretizing into finite elements or small areas where the solutions can be reduced to a system of algebraic equations. This is indicated in Figure 4–7.

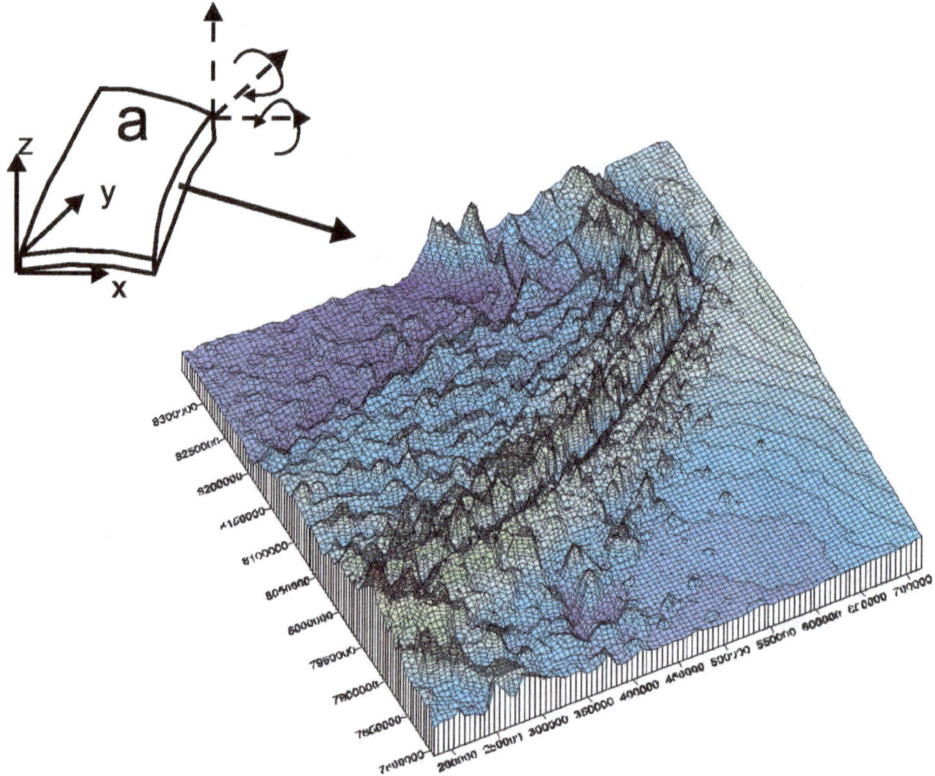

Figure 4-7 Discretisation of a finite element network; 'a' is the finite element, one of the small areas that build the model of Mohns Ridge. Click here for larger image.

In the modelling phase, the stress and displacement fields are modelled inside every finite element that builds the model.

Figure 4–8 below shows a 3D display of the deflection function used in the calculations. This is compared to the 3D bathymetric display of the same area in Figure 4–9. This similarity shows that the deflection matrix achieved by the FEM analyses compares quite well with the actual relief of the area under study, and confirms the strong link between the relief and the calculated crustal stresses using the chosen model of a 5 kilometer thick load.

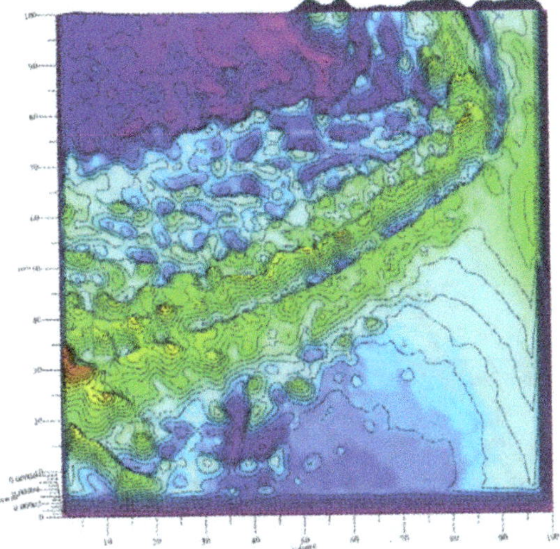

Figure 4-8 Relief of the deflection function.
Click here for larger image.

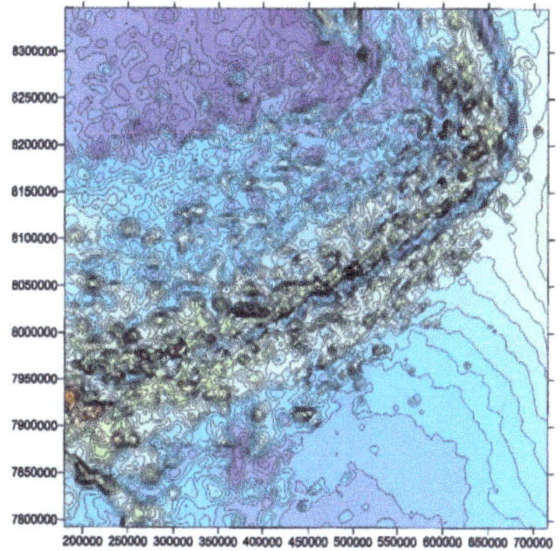

Figure 4-9 Multibeam relief.
Click here for larger image.

Knowing the deflection matrix, it is possible to define the inner forces acting within the plate and estimate the stresses.

Tests have demonstrated that the zero tangential stress isoline ($\tau_t = 0$) corresponds well with known active hydrothermal areas. Further tests have shown that relict and low-active hydrothermal manifestations are bound to the tensional areas where tangential stresses are larger than zero ($\tau_t > 0$) (Petukhov et al., 2004, 2010 and 2012, and Aleksandrov et al., 2009, in Cherkashov et al., 2013). See Figure 4–10.

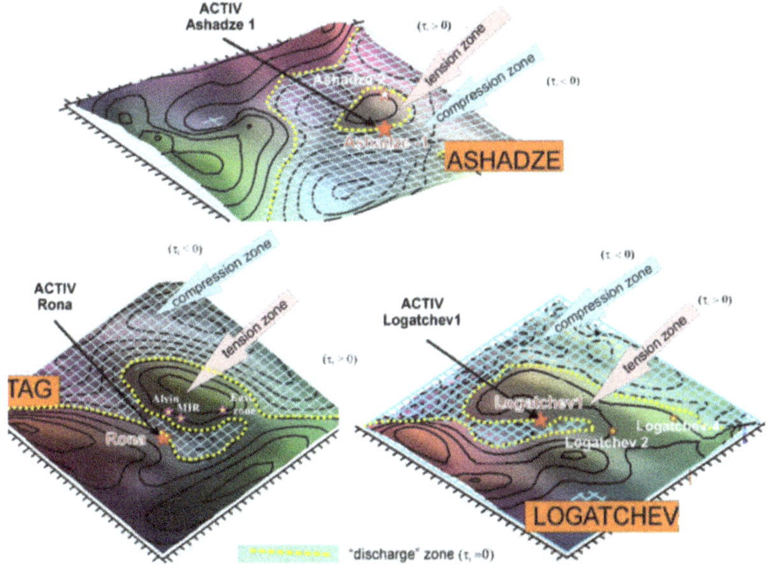

Figure 4-10 Models of tangential stress distribution MAR sections (Petukhov et al., 2004, 2010 and 2012, and Aleksandrov et al., 2009, in Cherkashov et al., 2013 . Illustration reproduced by permission; no reuse without rightsholder permission.).
a) Ashadze, b) TAG, c) Logatchev. The active fields are marked by a red star.
Click here for larger image.

51

The results presented for Ashadze, TAG and Logatchev in Figure 4–10 are crucial for the analysis of the deflection matrix and its interpretation. The results of calculating the tangential stresses in Mohns Ridge are shown in Figure 4–11.

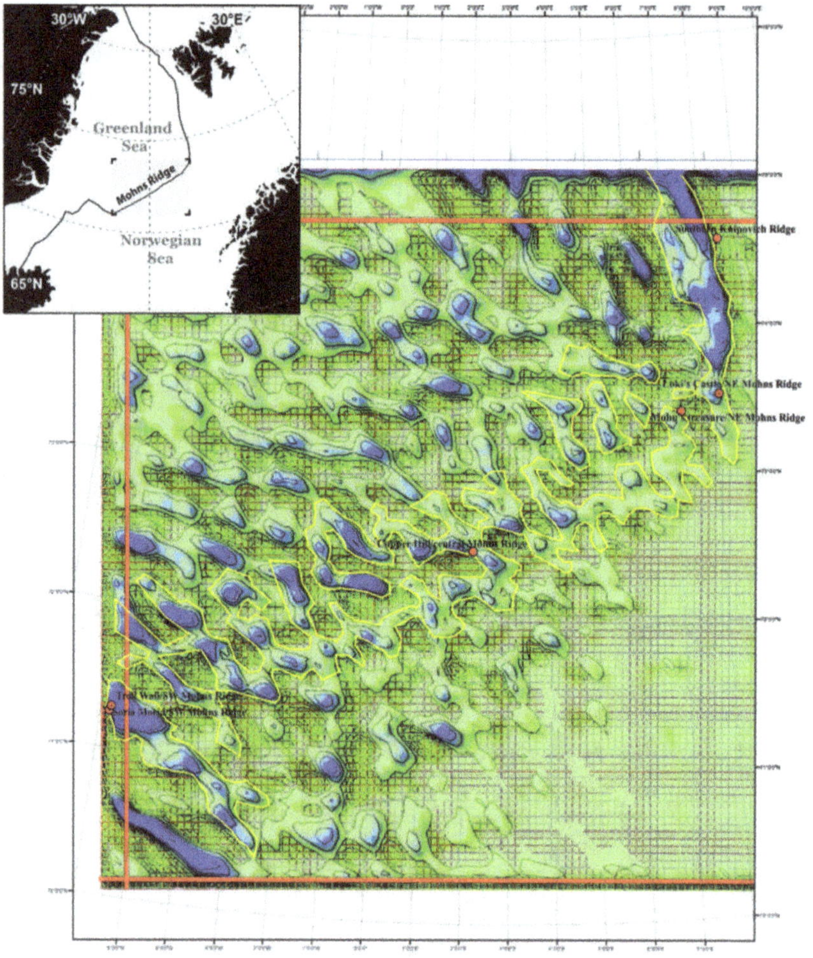

Figure 4-11 The distribution of tangential stress at the plate surface along Mohns Ridge. Yellow line is the zero stress isoline. Red dots are known hydrothermal manifestations (inactive and active) along Mohns Ridge (see Pedersen et al., 2010b). Click here for larger image.

A line with zero tangential stress can be plotted on the bathymetric map (yellow line in Figure 4–12 and Figure 4–13) that marks zones of possible discharge where stress changes from tension to compression. It is hypothesised that these zones are favourable to the presence of hydrothermal vents and match with the known hydrothermal sites along the Knipovich and Mohns Ridges shown in Figure 4–11, Figure 4–12, Figure 4–13, Figure 4–16 and Figure 4–17. Red lines in these figures indicate zones with a high concentration of favourable indicators for hydrothermal manifestation from the geodynamical analysis.

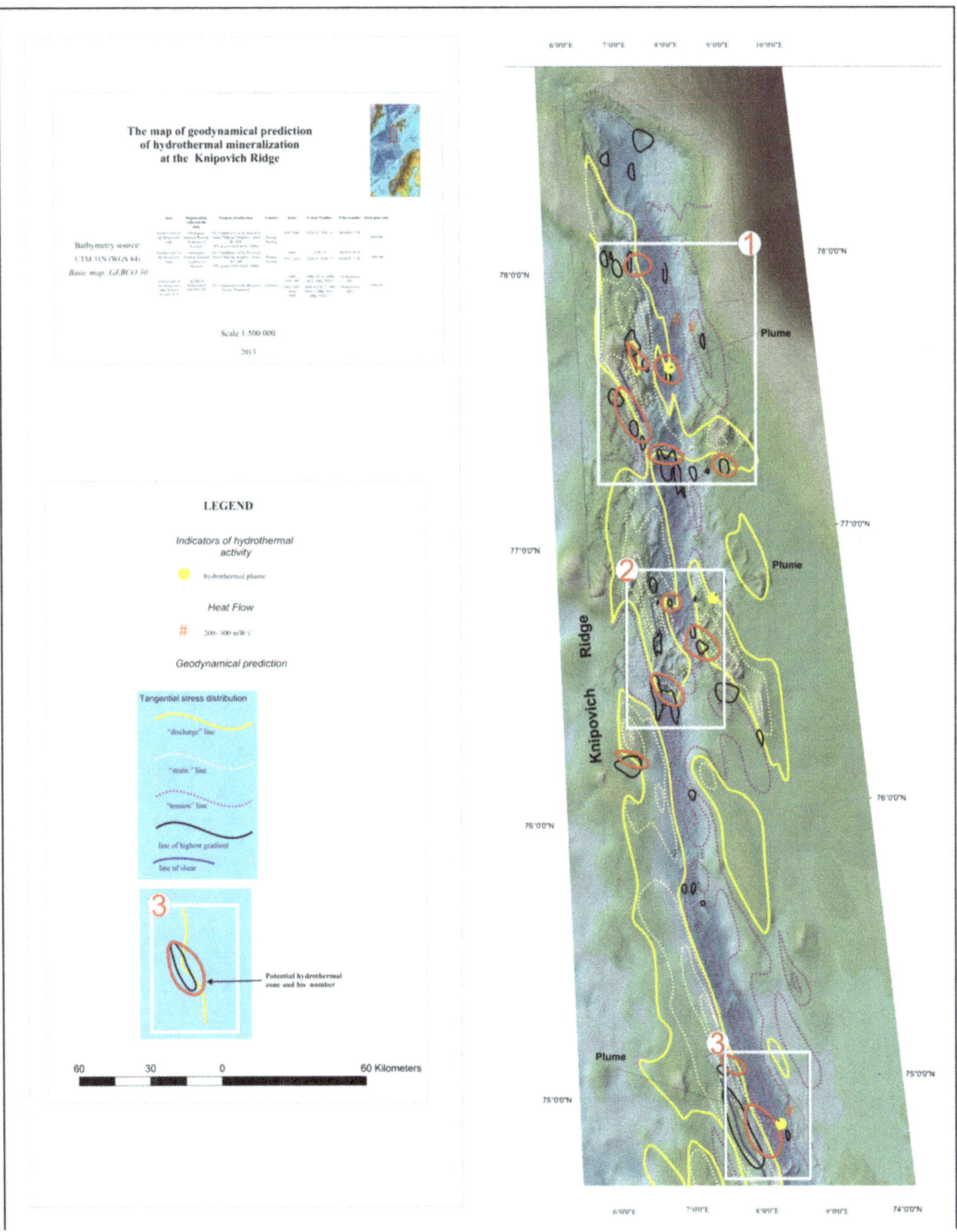

Figure 4–12 Areas most promising for finding SMS along the Knipovich Ridge, based on the geodynamic analysis. Areas having a concentration of prospective zones are shown as white boxes. These three areas (white box outline) and sub-areas have been modelled and named as separate plays in the play-database. Red lines indicate particularly promising areas, where the tensional structures can be found in close proximity to discharge zones and high gradients. Click here for larger image.

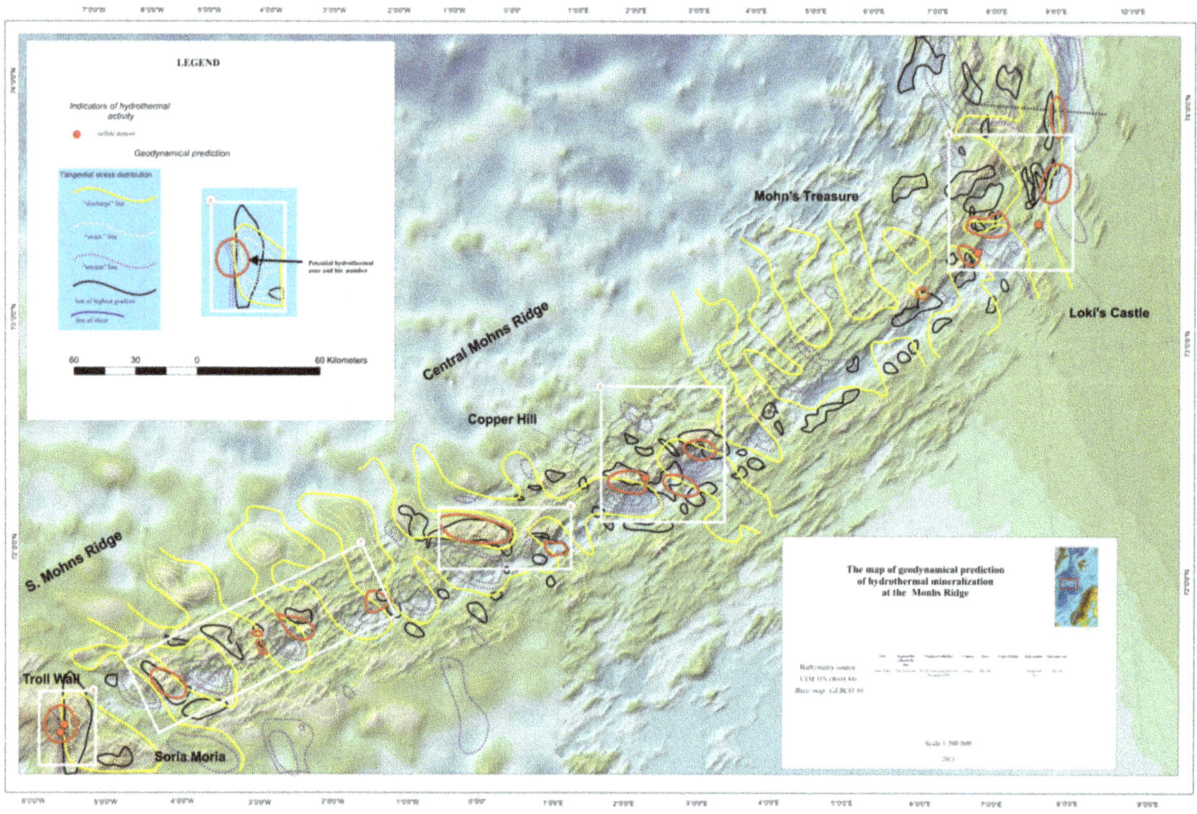

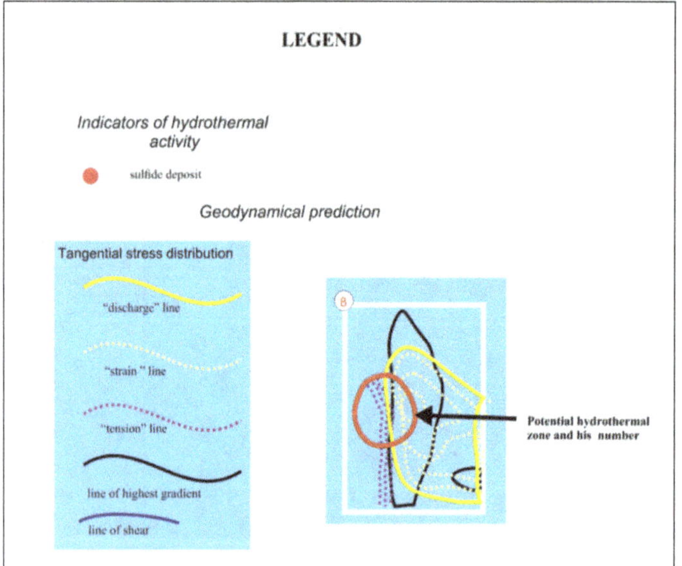

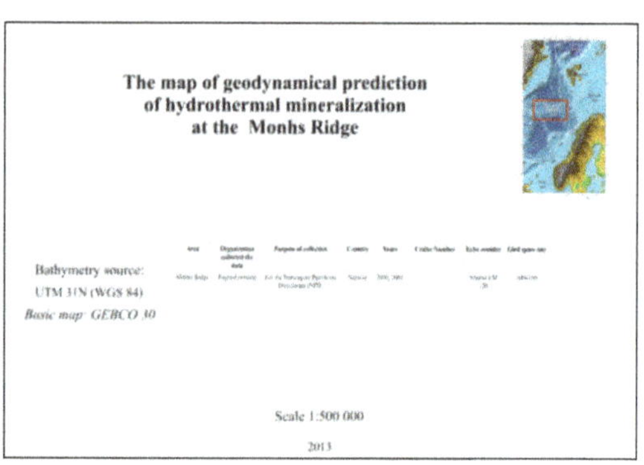

Figure 4-13 Areas most promising for finding SMS along Mohns Ridge, based on the geodynamical analysis. Areas having a concentration of prospective zones are shown as white boxes. These five areas (white box outline) and sub-areas have been modelled and named as separate plays in the play-database. Red lines indicate particularly promising areas, where the tensional structures can be found in close proximity to discharge zones and high gradients. Click here for larger image.

4.1.4 Morphostructural analysis

4.1.4.1 Data and methodology

Mohns Ridge

Morphostructural analysis of Mohns Ridge was based on the bathymetric surveys carried out by Fugro-Geoteam for the Norwegian Petroleum Directorate (NPD) in 2000 and 2001.

An original grid (100 x 100 m) was built using mean values of the primary data, which allowed locating details of the relief. Several bathymetric charts at different scales were drawn. However, the lack of a sonar survey has limited the accuracy of interpretations.

The morphostructural scheme of the axial zone of Mohns Ridge was drawn with the aim of determining tectonic and volcanic elements of the seabed structure, comprising both large and small tectonic and volcanic structures. Unfortunately, due to the absence of sonar images, it was not possible to differentiate between the seabed volcanic formations at the bottom of the valley by their age. However, it was possible to isolate structures associated with the detachment faults and oceanic core complexes (OCC).

Regarding the selection of promising areas for sulphide ore formation, general criteria were followed. All known vents on the ridges with slow or ultra-slow spreading are generally attributed to the axial or flank features (Krasnov et al., 1995). The former is directly controlled by the processes of axial volcanism. This could either be a thermal one, caused by the presence of a high-temperature magmatic heat source at shallow depths under the seabed or by fluids accompanying volcanic outflows from a deep source. The latter, on the other hand, do not demonstrate any noticeable association with the volcanic processes, although an indirect association is implied and is located at remote flanks or edges of the rift valley, mainly in the areas of the development of serpentinised peridotites.

In the majority of cases, the axial venting is concentrated at small grabens, which complicates apical surfaces of the axial volcanic ridges at their most elevated parts. The SMS deposits normally tend towards the areas of very young basalts in the zones of intersection of the longitudinal faults and the small transversal dislocations with the indications of tectonic tension at areas with tangential stresses close to zero. Large, central-type volcanoes may also serve as ore-controlling structures. This is valid in cases where they are located in the central parts of the inner floor of the rift valley and areas where they lie at the toes of the magma-conducting marginal faults having tangential stresses close to zero.

The flank SMS deposits are structurally controlled by zones of transection of deep-seated longitudinal strike-slip faults, mainly the detachment faults, with crossing tectonic dislocations. These provide an additional possibility for lateral subsurface migration of magmatic and hydrothermal fluids and, on some occasions, even for the melts far beyond the boundaries of the axial zone. In general, the ore-localising structures are represented by the flattened slopes, inter-ridge depressions and saddles. Recent research has proved that there exist quite high promises of ore-bearing OCC. This is especially the case when these are located immediately close to the axial zone (Cherkashov et al., 2013). One may notice that Mohns Ridge, from a morphostructural perspective, more or less meets the above-mentioned criteria.

The areas are outlined in such a way that the most promising structures are grouped together on the basis of the following:

- the complex tectonic areas with volcanic ridges having the axial grabens. In particular, with the volcanic steps and links/bridges connecting the volcanic rises and slopes.
- the structures spatially or generically associated with the detachment faults and OCC at the valley walls/slopes, especially areas where tensional environment exists due to close to zero tangential stresses.
- the large central-type volcanoes with the flat tops and craters and the chains of such volcanoes along the crossing faults with close-to-zero tangential stress environment.

Basically, the areas include axial and adjacent flank structures. However, in some cases, several remote flanks have also been included since their structures suggest the existence of extinct/relict vents.

Figure 4–14 shows the morphostructural elements along Mohns Ridge and Figure 4–16 show areas most promising for finding SMS deposits based on the morphostructural analysis.

Knipovich Ridge

Morphostructural analysis of the Knipovich Ridge axial zone has been made using the data from several bathymetric surveys. A summary bathymetric map has been drawn based on the original (R/V Haakon Mosby and R/V Polarstern data) and available (R/V Nikolai Strakhov data) grids. The general quality of the primary materials appeared to be poorer than that of Mohns Ridge, which limited the extent of details for interpretation. This drawback was partly compensated for by using side-scan sonar images (SEAMARC-II and ORETECH systems; (Crane et al., 2001.), which allowed for the outlining of areas of volcanic activity over time, as well as the zones of recent tectonic activity in the northern part of the ridge.

Echo-sounder survey data confirmed by video (Crane et al., 1995, 2001; Tamaki & Cherkashev, 2000) manifests that there have been recent volcanic activities on the major part of Knipovich Ridge, excluding its two sections in the interval of 74 to 75 degrees north and north of 78 degrees. One may notice that both sections are more skewed to the theoretical axis of spreading, where an increase in sedimentary cover of up to 2 km is recorded. The ORETECH materials (Crane et al., 2001) demonstrate somewhat lower volcanism in the southern half of the ridge. As the survey covered a relatively narrow strip (only 2 km) of the rift valley, this conclusion is uncertain.

Generally, hydrothermal manifestations and associated sulphide accumulations at Knipovich Ridge may be associated with several areas of modern tectonic and volcanic activities in the zones of intersection of multi-directional faults and structural features where tangential stresses are close to zero.

The morphostructural scheme of the Knipovich Ridge is based on the same principles and the same legend as that for the Mohns Ridge scheme. In addition, the northern half consists of fresh and older sheet and pillow lavas, as well as large elevated massifs at the flanks of the valley.

Figure 4–15 shows the morphostructural elements of the Knipovich Ridge and Figure 4–17 show areas most promising for finding SMS deposits, based on the morphostructural analysis along Knipovich Ridge.

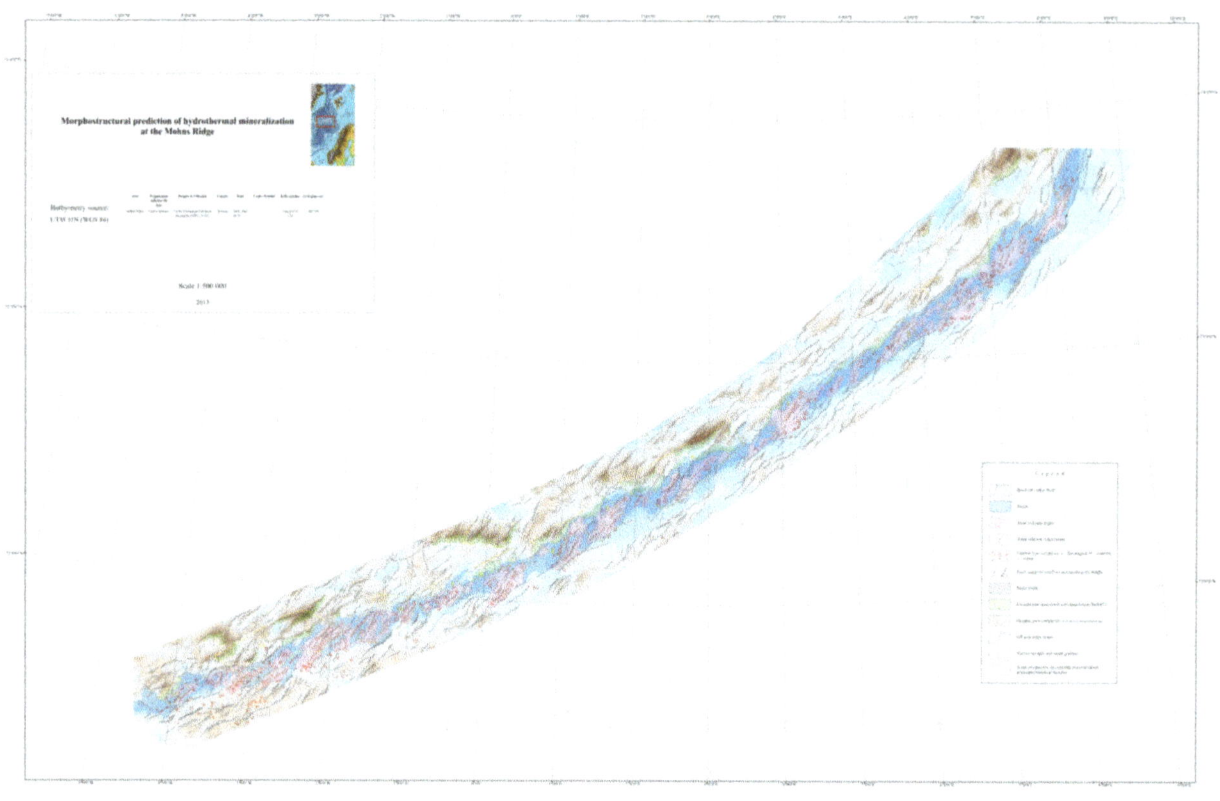

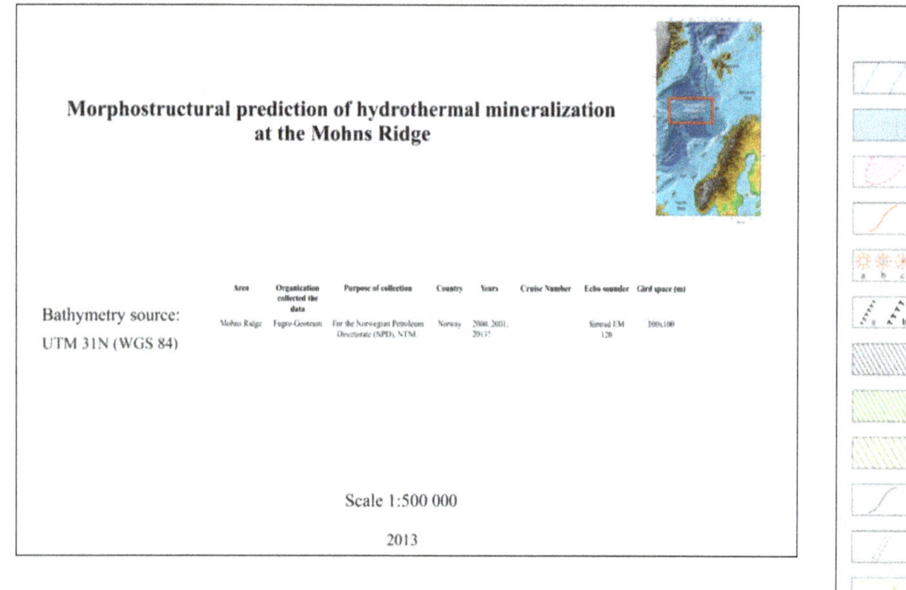

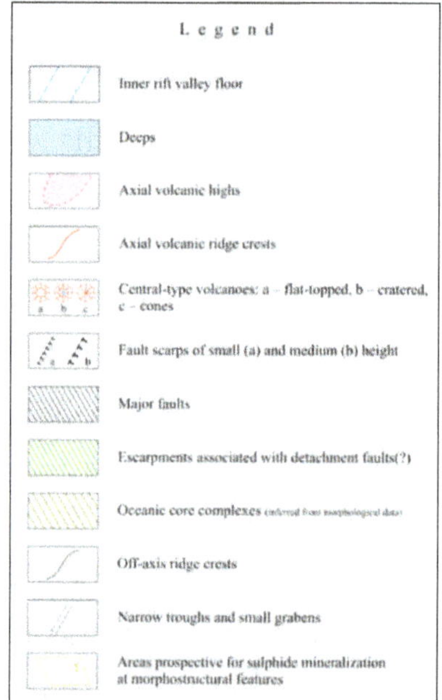

Figure 4-14 Morphostructural elements along Mohns Ridge. Click here for larger image.

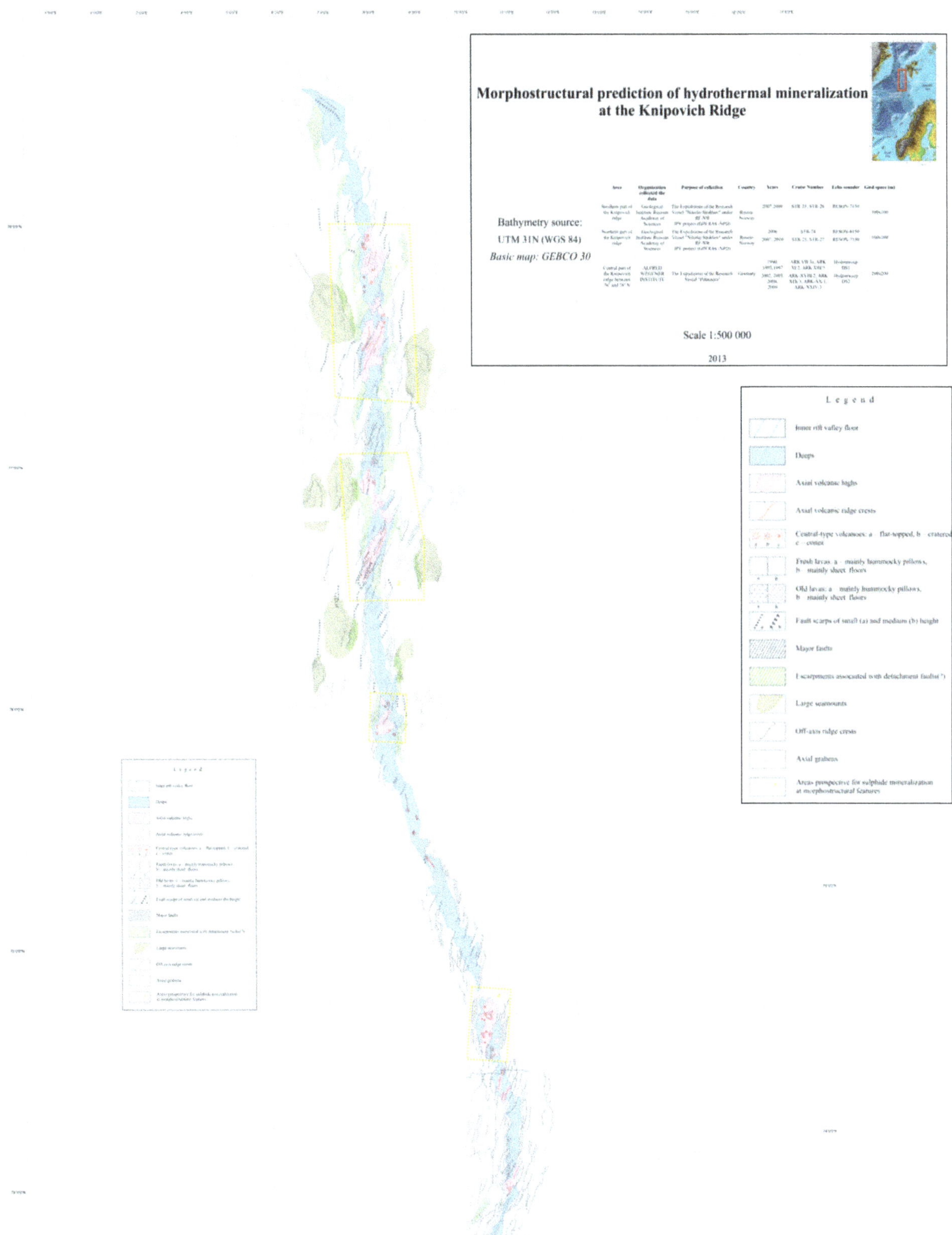

Figure 4-15 Morphostructural elements along Knipovich Ridge. Click here for larger image.

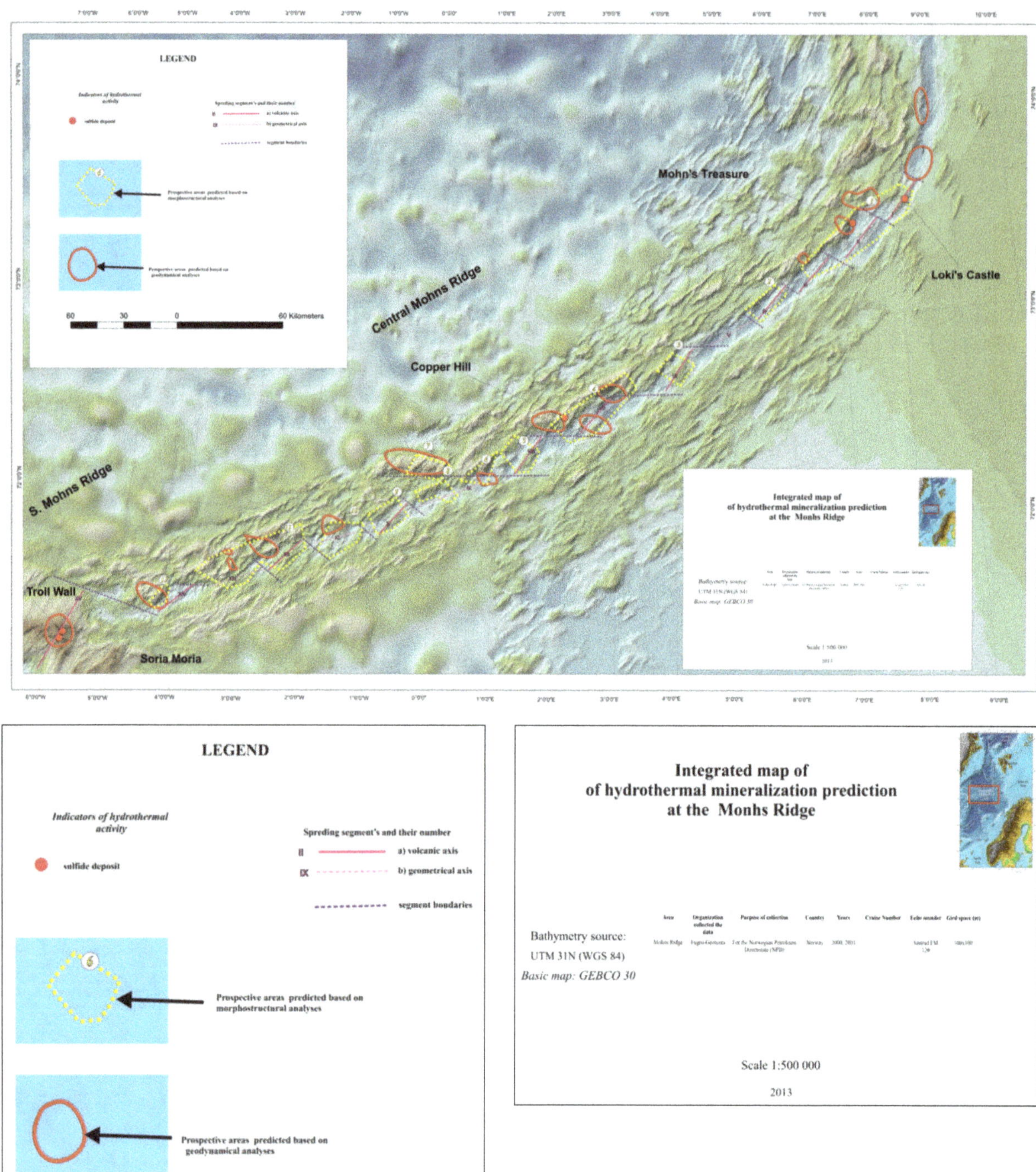

Figure 4-16 Areas most promising for finding SMS along Mohns Ridge, based on the morphostructural analysis. Segments also indicated. Areas having a concentration of prospective zones are shown as yellow dotted shapes. Click here for larger image.

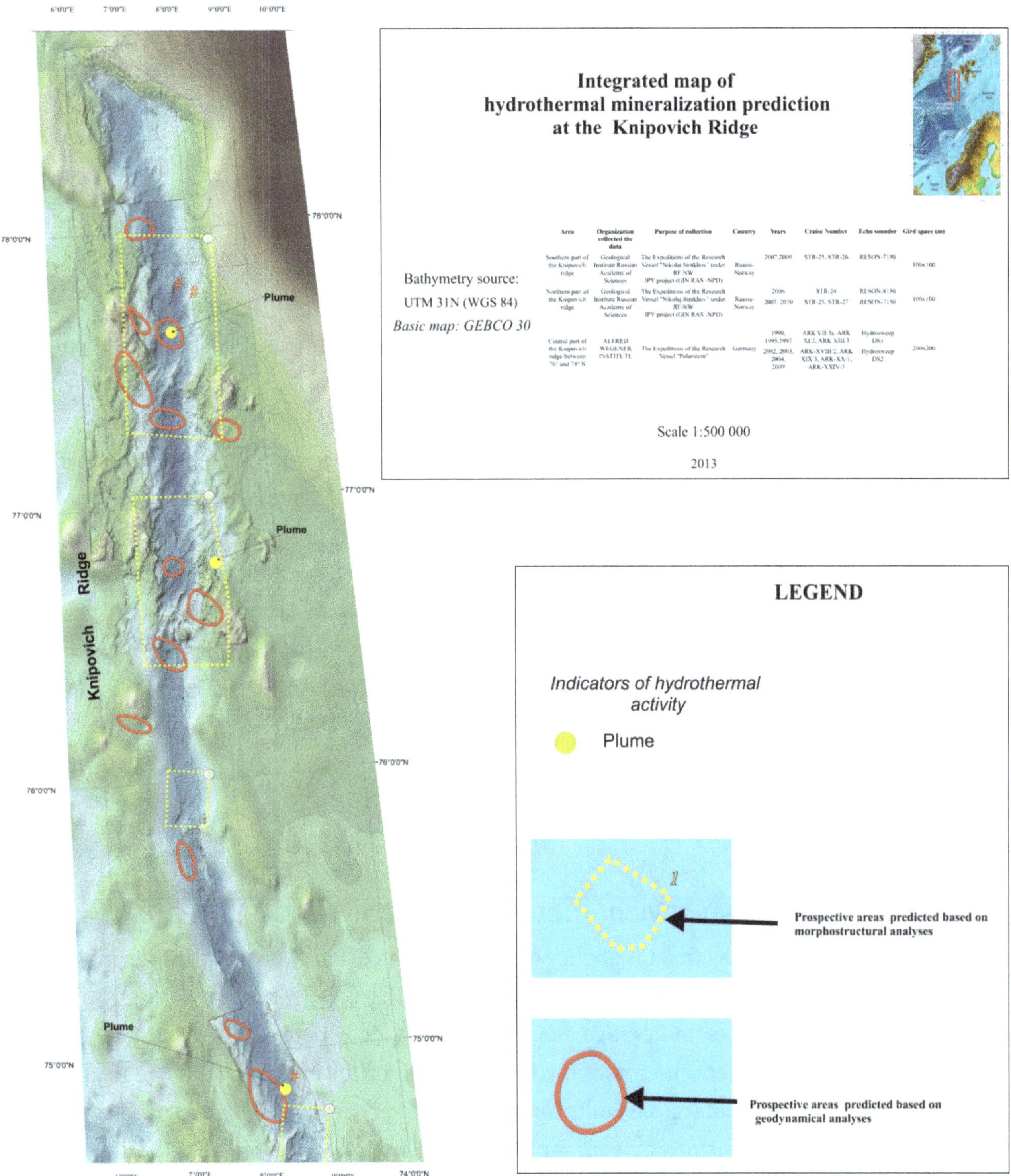

Figure 4-17 Areas most promising for finding SMS along the Knipovich Ridge, based on the morphostructural analysis. Areas having a concentration of prospective zones are shown as yellow dotted shapes. Click here for larger image.

4.2 Descriptions of single permissive tracts

4.2.1 Introduction

This sub-chapter contains a detailed description of the different permissive tracts that have been defined based on morphostructural and geodynamical analysis. Mohns Ridge has been segmented as shown in the previous sub-chapters. The descriptions are based on Cherkashov et al. (2013). The descriptions form the formal basis for the permissive tract characterisation in Chapter 6.

In total, eight areas have been defined and delineated based on the geodynamic analysis along Knipovich (3) and Mohns Ridges (5). These are numbered and outlined in white rectangles in Figure 4–12 and Figure 4–13 in the previous sub-chapter. From the morphostructural analysis, 16 favourable areas (MS-areas) have been identified, and these are indicated by yellow dotted lines in Figures 4–14 and 4–15. The permissive tracts are geographically centred on these 16 areas and around the southernmost part of Mohns Ridge, containing the Soria Moria and Troll Wall vent fields. The size of respective tracts in square kilometres has been calculated based on these yellow MS-areas. Favourable areas on Knipovich Ridge have been given these IDs: MS-K-area 1, MS-K-area 2 etc.; whereas favourable areas on Mohns Ridge have been given these IDs: MS-M-area 1, MS-M-area 2 etc.

4.2.2 Knipovich – permissive tract K1 and MS-K-area 1

The MS-K-area 1 is about 80 km long and covers two spreading segments of the rift valley, including its adjacent flanks. See Figure 4–18. A large volcanic-tectonic ridge, 600 metres high, dominates the southern part of the segment. It crosses the southern end of the area at a small angle and joins the marginal faults forming the ridge axis flanks. The surface of the ridge is formed by young pillow lavas, and a central graben is seen at its most elevated part. The slopes are controlled by tectonic scarps, spatially and perhaps genetically associated with the high-amplitude faults (possible detachment faults) of the valley walls. West of the axial ridge, a much smaller extinct volcanic ridge can be observed.

In its central part, the axial valley floor is largely covered with sediments and the lavas are seen only on a very small volcanic rises and in a strip to the north.

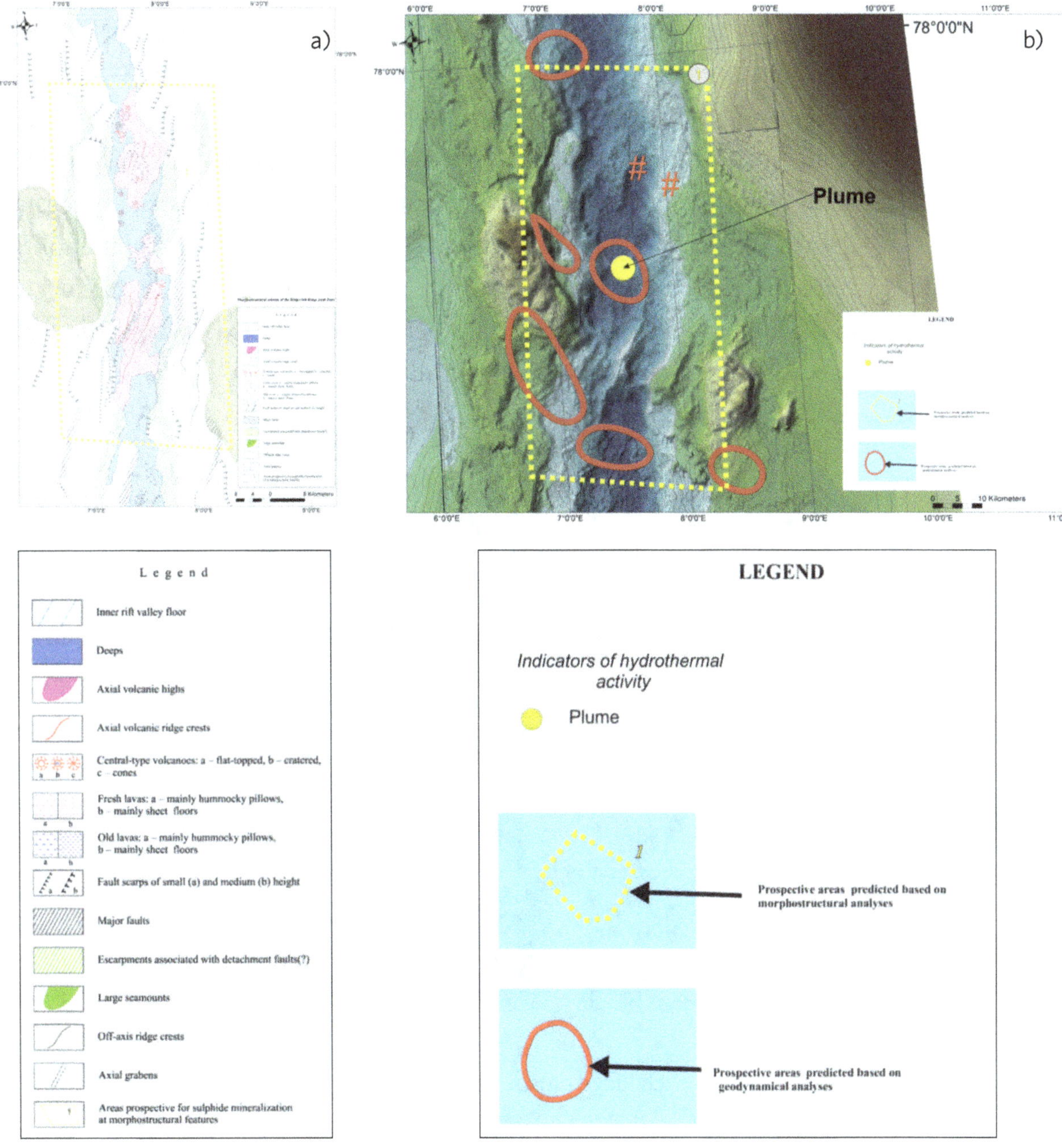

Figure 4–18 Knipovich Ridge interest area 1. Yellow dotted line is the area defined in the morphostructural analysis. Figure a shows the morphostructural elements, whereas figure b shows the prospective areas identified by the geodynamic analysis outlined in red. Red hashtags in b indicate the location of heat-flow measurements between 200 and 300 mW/m². Click here and here for larger image.

In the northern part of the area, there is also a large irregular volcanic rise with a relative height of 500 metres, with basalts of different ages. To the north of this rise, a cluster of central-type volcanoes, the largest being more than 3 kilometres in diameter, can be observed. An increased heat flow of up to 300 mW/m² is reported at two locations around 77048'N (Zayonchek et al., 2011 in Cherkashov et al., 2013).

The walls of the rift valley are mainly controlled by the large faults that could be interpreted as detachment faults. Of particular interest is a large zone of transverse tectonic deformations of north-west strike, crossing the rift valley at latitudes corresponding to the central parts of the southern axial ridge. On the flanks of the valley, several very large, elevated massifs of complex and uncertain genesis are located. Two of these are partly included in the area under study.

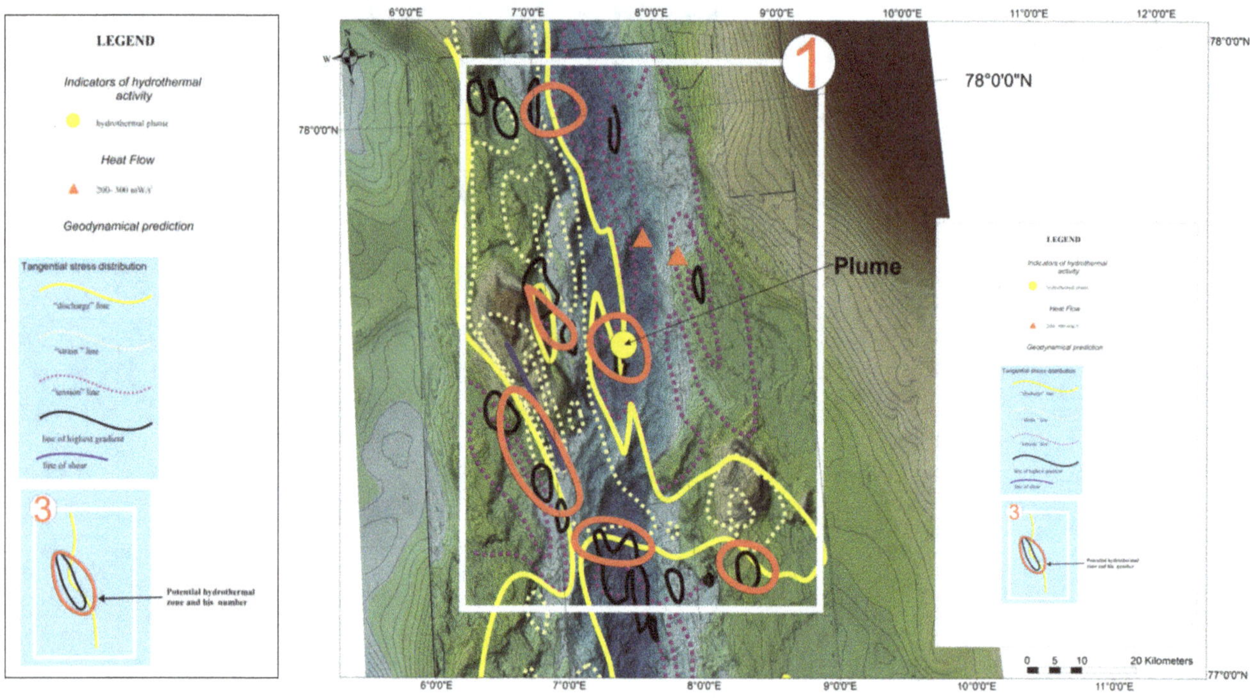

Figure 4-19 Knipovich Ridge prospectivity of interest from the geodynamical analysis in Area 1. Click here for larger image.

Area 1 includes several promising sites (outlined in red in Figure 4–19). In one of them, a hydrothermal plume has been recorded at 77°40′ N (Beaulieu, 2015). In area 1, an increased seafloor heat flow, of up to 300 mW/m², at 77°48′ N has been recorded and reported (Zayonchek et al., 2011 in Cherkashov et al., 2013).

By morphostructural analysis, the potential for finding SMS of an axial type is connected mainly with the central graben of the southern axial ridge and the volcanic cluster in the north. The location of the off-axial hydrothermal manifestations that one may expect will primarily be near the junction between volcanic structures with marginal faults, as well as along the tectonic dislocation zones, including the slopes of large flange massifs.

4.2.3 Knipovich – permissive tract K2 and MS-K-area 2

The MS-K-area 2 includes a section of the rift valley about 65 km long inside two adjacent spreading segments. See Figure 4–20 and Figure 4–21. The structure plane of its southern half is defined by the 500 m linear volcanic ridge, in many respects similar to the one described earlier, which at a small angle transects the inner bottom of the rift valley and merges with the marginal scarp of the western valley wall northwards of the zone of non-transform offset. According to the side-scan sonar data, the ridge has a multi-aged structure. The youngest part is practically uncovered by sediment. Basalts are found in the lower part of the eastern slope of the ridge and in the adjacent linear depression.

In the central part, the most interesting aspect is a bathymetric rise located in the zone of the non-transform discontinuity; it is a 120 m transverse terrace of volcanic origin, and its surface consists of very young pillow and sheet basalts without any noticeable sedimentary overburden. The south-east slope of the terrace is controlled by a tectonic scarp parallel to the spreading axis, and at this toe, young lava breccia have developed. Observations from MIR submersible dives here showed a very dense net of fractures in three main directions (N = North, E=East) NoE – N10E degrees, N30E – N40E degrees and N120E – N130E degrees. The other volcanic rises are relict structures.

a)

b)

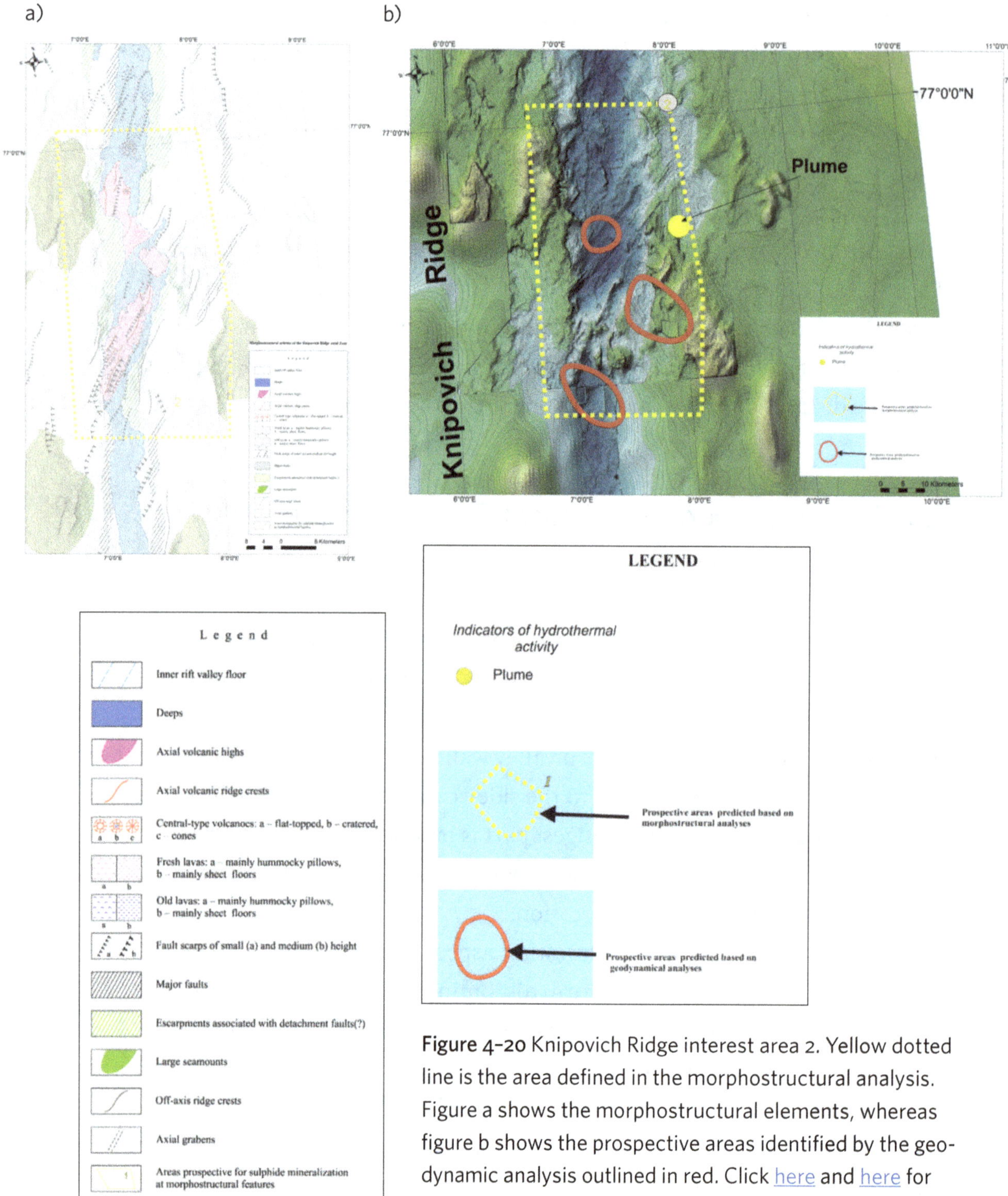

Figure 4-20 Knipovich Ridge interest area 2. Yellow dotted line is the area defined in the morphostructural analysis. Figure a shows the morphostructural elements, whereas figure b shows the prospective areas identified by the geodynamic analysis outlined in red. Click here and here for larger image.

Two 200 m central-type cratered volcanoes, 2 and 4 km in diameter, are located near the northern boundary. They are connected with the development of young sheet basalt lavas.

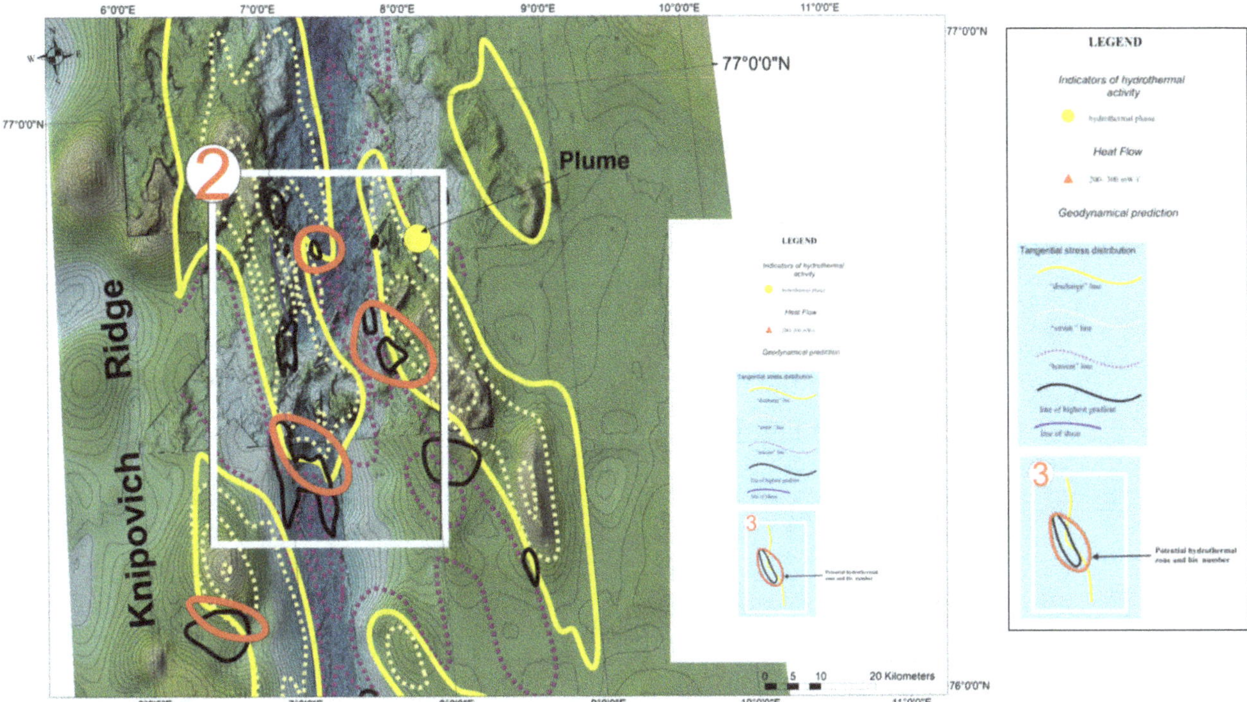

Figure 4-21 Knipovich Ridge prospectivity of interest from the geodynamical analysis in Area 2. Click here for larger image.

The prospect of finding sulphide mineralisations of an axial type here is connected mainly with the central graben of the southern axial ridge and the volcanic cluster in the north, similar to that described for Area 1. The inner rift-valley walls are mostly associated with high-amplitude faults (detachment faults).

Several promising areas have been found in Area 2 by the geodynamic analysis (indicated by red lines). All of them are discharged (indicated by the yellow line) and are situated near the high-gradient tension structures (indicated by black lines). See Figure 4–21. A plume has been detected in Area 2 at 76°48′ N (Beaulieu, 2015). This is located in the zone defined as a discharge zone (outlined in yellow).

The most promising areas are the axial volcanic ridge, the volcanic terrace, the valley slopes close to transverse tectonic disturbances, and in the zones of the junction between these transverse disturbances and the axial volcanic structures. See Figure 4–20.

4.2.4 Knipovich – permissive tract K3 and MS-K-area 3

MS-K-area 3 includes a 20 km section of the rift valley, located in an area of minor non-transform offset, marked by a 400 m axial volcanic rise and several flat-topped volcanoes of the central type. See Figure 4–22.

The tract is defined solely through the morphostructural analysis.

Young basalts are widely spread in this area (see also Crane et al., 2001). One may suppose, with a certain level of probability, that modern or extinct vents may exist in the axial zone and neighbouring flanges associated with the detachment faults interpreted in the area (see green shaded areas in Figure 4–22a).

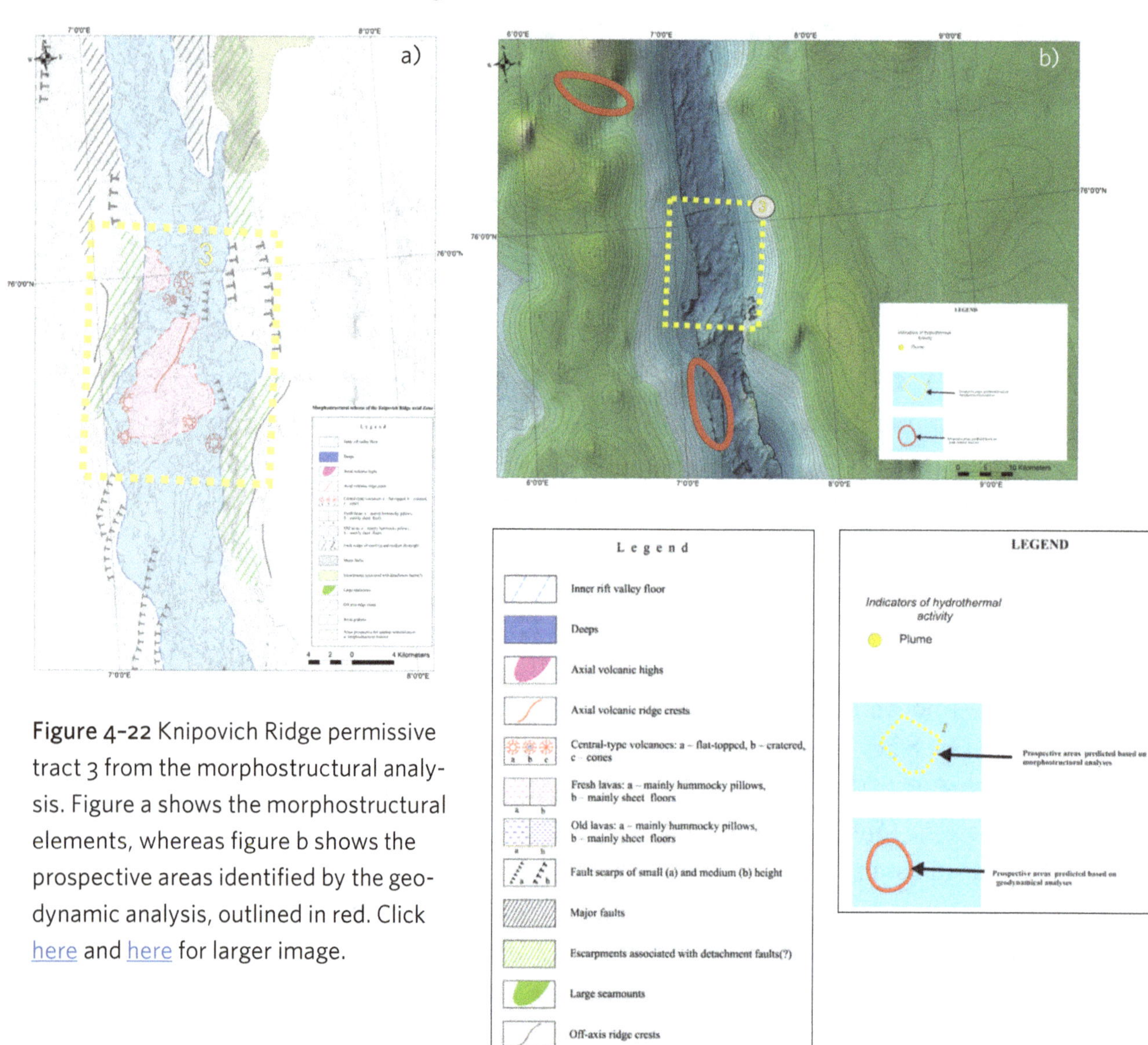

Figure 4-22 Knipovich Ridge permissive tract 3 from the morphostructural analysis. Figure a shows the morphostructural elements, whereas figure b shows the prospective areas identified by the geodynamic analysis, outlined in red. Click here and here for larger image.

4.2.5 Knipovich – permissive tract K4 and MS-K-area 4

MS-K-area 4 is connected to Area 3 from the geodynamic analysis (see Figure 4–23). It contains promising sites. There are hydrothermal plumes at 74°48′ N and 74°53′ N (Beaulieu, 2015). Although the plume at 74°53′ N is not confirmed, the nearby station (8°30′ E, 74°53′ N) has picked up a heat-flow measurement of 233 mW/m² (Zayonchek et al., 2010). Only the southernmost plume is shown in Figure 4–23.

Both hydrothermal plumes are in the estimated zone of discharge (indicated by the yellow line). However, due to uncertainties both with regard to the source of the plume and the position of the estimated zone of discharge, nothing conclusive can be stated. In Area 3, promising specific sites in addition to the plumes have been identified. They are located in the discharge zone close to the tension structure (western flank).

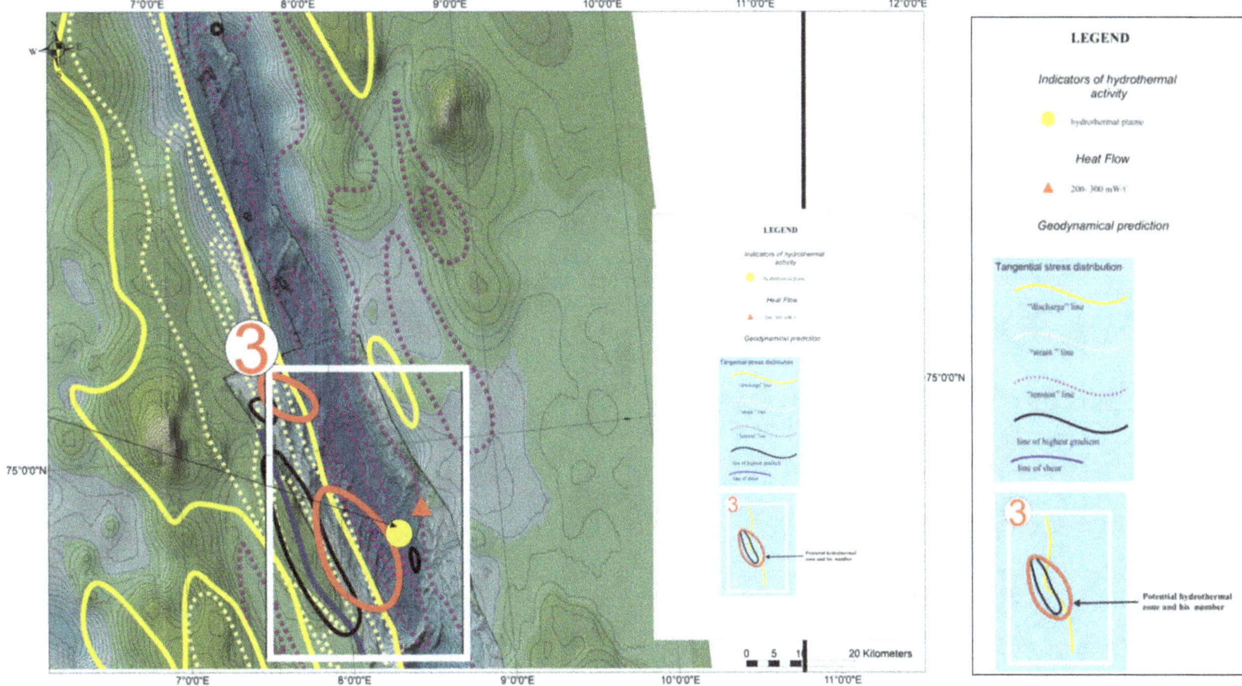

Figure 4–23 Knipovich Ridge prospectivity of interest area 3, defined by the geodynamical analysis. Orange triangle indicates the heat-flow anomaly. Click here for larger image.

a)

b)

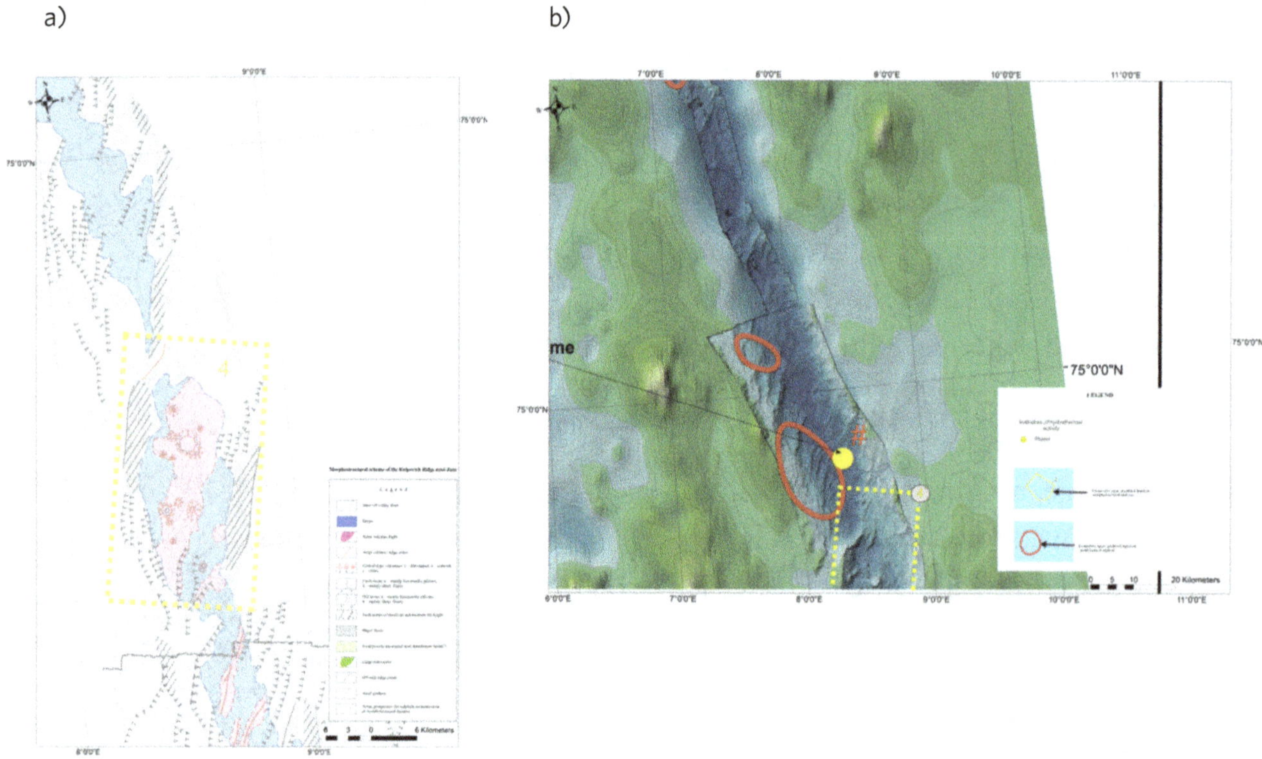

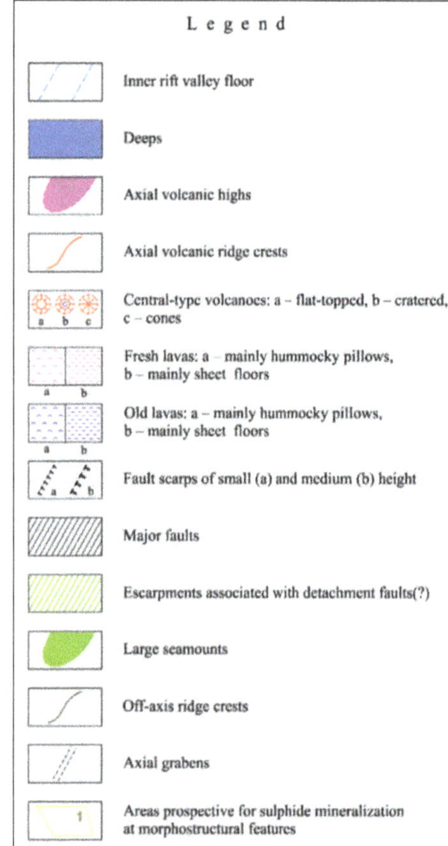

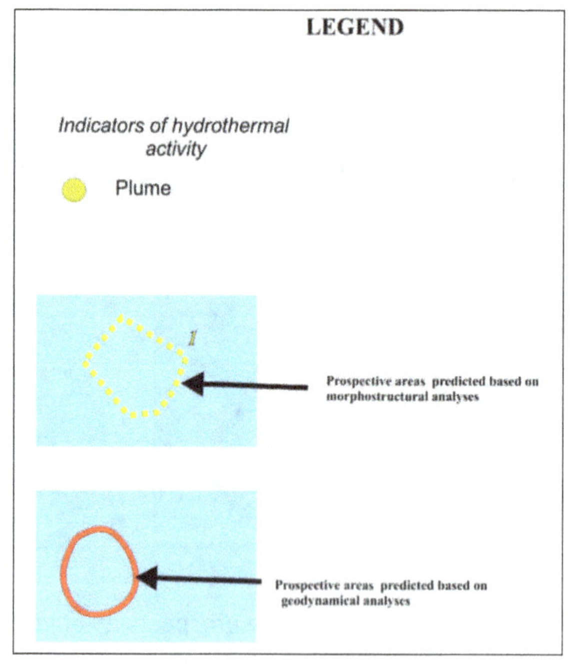

Figure 4-24 Knipovich Ridge permissive tract 4 from the morphostructural analysis. Figure a shows the morphostructural elements, whereas figure b shows the prospective areas identified by the geodynamic analysis, outlined in red. Click here and here for larger image.

In the region at 75.45° N. along the Knipovich Ridge, an abnormal turbidity was recorded at depths of 2600–2650 metres, along a 2 km zone (Cherkashov et al., 2013). Increased contents of Mn and dissolved CH4 were observed. At the northern extremity, the anomaly is at a depth of 2850 metres, and at 2650–2700 metres at the southern extremity. One may suggest that the source of the plume anomalies lies further to the north, where outcrops of ultrabasic lithologies / rock types are likely, including serpentinites.

Area 3 in Figure 4–23 overlaps partly with segment approximately 30 km long, with distinct rhomboid outlines. See Figure 4–24. These are bound by rather specific non-transform offsets, where an abrupt contraction of the inner floor occurs due to spatial convergence of the opposite scarps of the valley walls of the adjacent segments. A low (200 m) sub-meridional linear volcanic rise without a clear crest runs along the whole section, and contains multiple central-type volcanoes, the largest reaching 3 km in diameter. The axial peak seems to be a very young structure and its surface is partly overlapped with young sheet lavas (Crane et al., 2001).

In the southern part of the section, the western rift valley wall represents a system of small steps. The eastern wall contains a monolithich structure and a steep scarp.

The promise of the area is associated with the high-amplitude flank faults. Of special interest is a narrow, high, probably tectonic ridge, which crosses the rift valley in the zone of non-transform offset in the north of the area. Also, the axial ridge volcanoes are of interest.

4.2.6 Mohns – permissive tract M1 and MS-M-area 1

Several promising areas have been identified by both geodynamical and morphostructural analysis, in and around interest area 4 from the geodynamical analysis. They are in the discharged condition and lie near the tension structures. These areas are promising for hydrothermal activity.

Hydrothermal manifestations found earlier confirm the prospectivity of permissive tract M1 / area 4 (Pedersen et al., 2010b). Loki's Castle is located at 73°33′ N in the discharge zone (indicated by yellow lines in Figure 4–25).

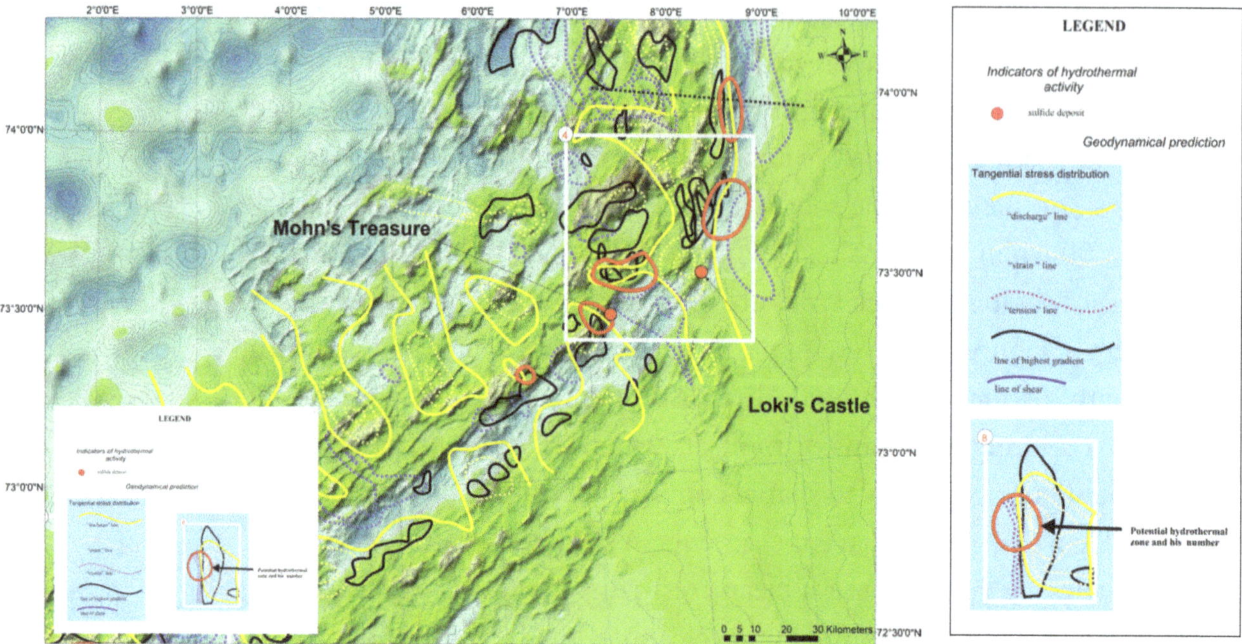

Figure 4-25 Mohns Ridge prospectivity of interest area 4, defined by the geodynamical analysis. Loki's Castle and Mohns' Treasure are both close to the yellow discharge line. Click here for larger image.

The second, Mohns' Treasure, at 73°27′ N is also in the discharged zone, and lies near the high-gradient tension structure. See again Figure 4–25. Their locations coincide with the areas identified as 'promising' by the geodynamical analysis, which is an indication of the appropriateness of the approach to revealing areas of hydrothermal activity.

MS-M-area M1 includes a section of the rift valley over 70 km long, having an almost complete set of morphostructural indicators for hydrothermal mineralisation (faults, central volcanos, scarps, axial volcanic ridges, graben structures etc.). This area has been divided into three segments. See Figure 4–26b. The most promising are the axial volcanic ridges with the central grabens and the multiple flat-topped and cratered central-type volcanoes, the rift-valley wall structures defined by high-amplitude faults (in particular, detachment faults) and the oceanic core complexes defined based on their morphology characterised by break-away ridges and corrugated surfaces. Thus, in the north-east, beside the known vent site on the axial graben (Pedersen et al., 2010b), one may assume a high probability for recent venting in the area, with transverse volcanic terraces at the toe of the eastern marginal scarp of the rift valley. Of special interest, is the western rift-valley wall in the middle and southern parts of the area, as well as a large central-type volcano at the foot of the eastern wall.

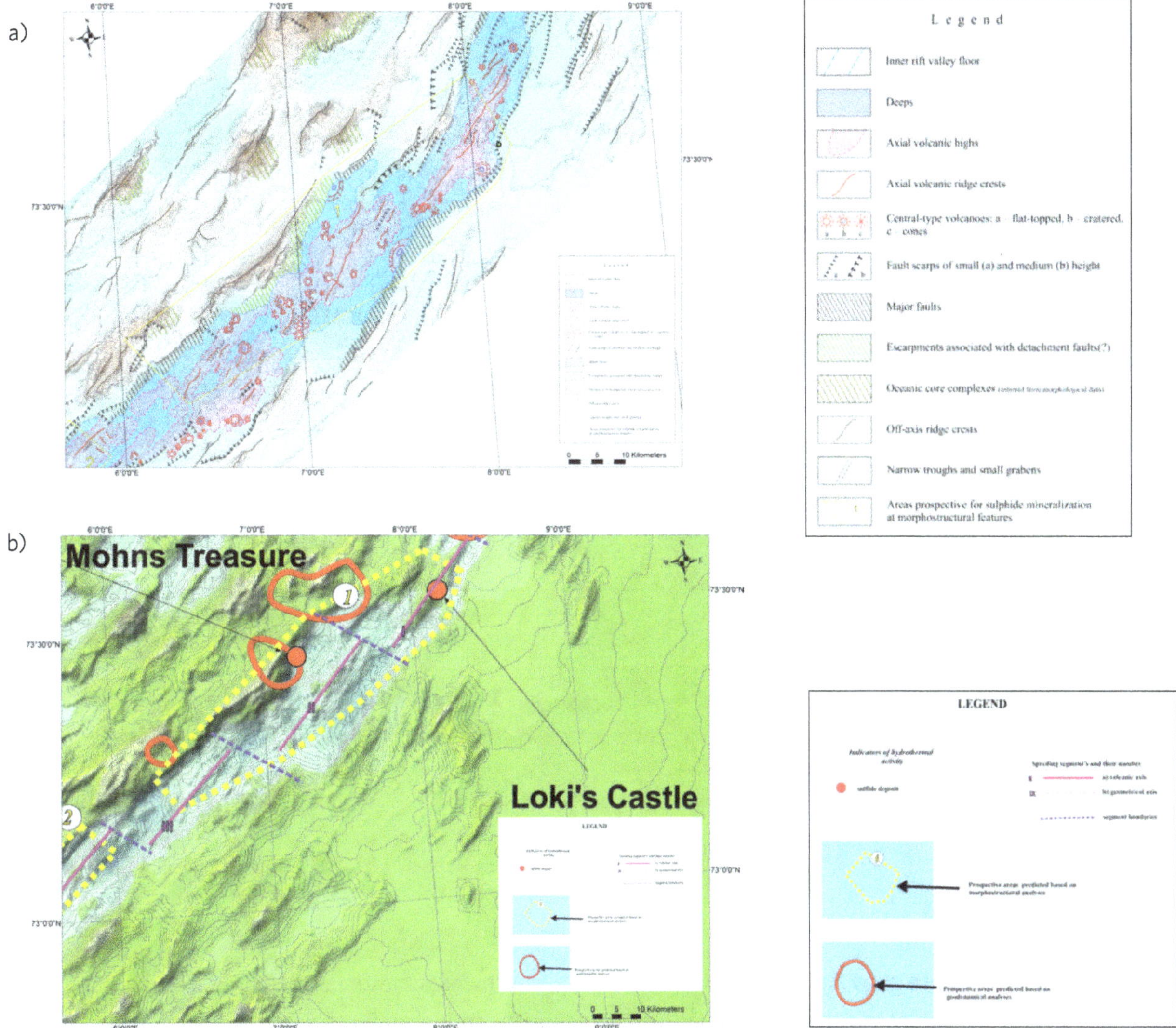

Figure 4–26 Mohns Ridge MS-M-area 1, close to interest area 4 defined by the geodynamical analysis. Figure a shows the morphostructural elements, whereas figure b shows the prospective areas identified by the geodynamic analysis, outlined in red. Click here and here for larger image.

4.2.7 Mohns – permissive tract M2 and MS-M-area 2

The MS-M-area 2 is in principal similar to the MS-M-area 1 described in the previous section. It includes a considerable part of segment IV, a rather high (700 m) axial volcanic ridge with the central graben in its most elevated part, and the western flank of the rift valley partly controlled by a detachment fault. See Figure 4–27a.

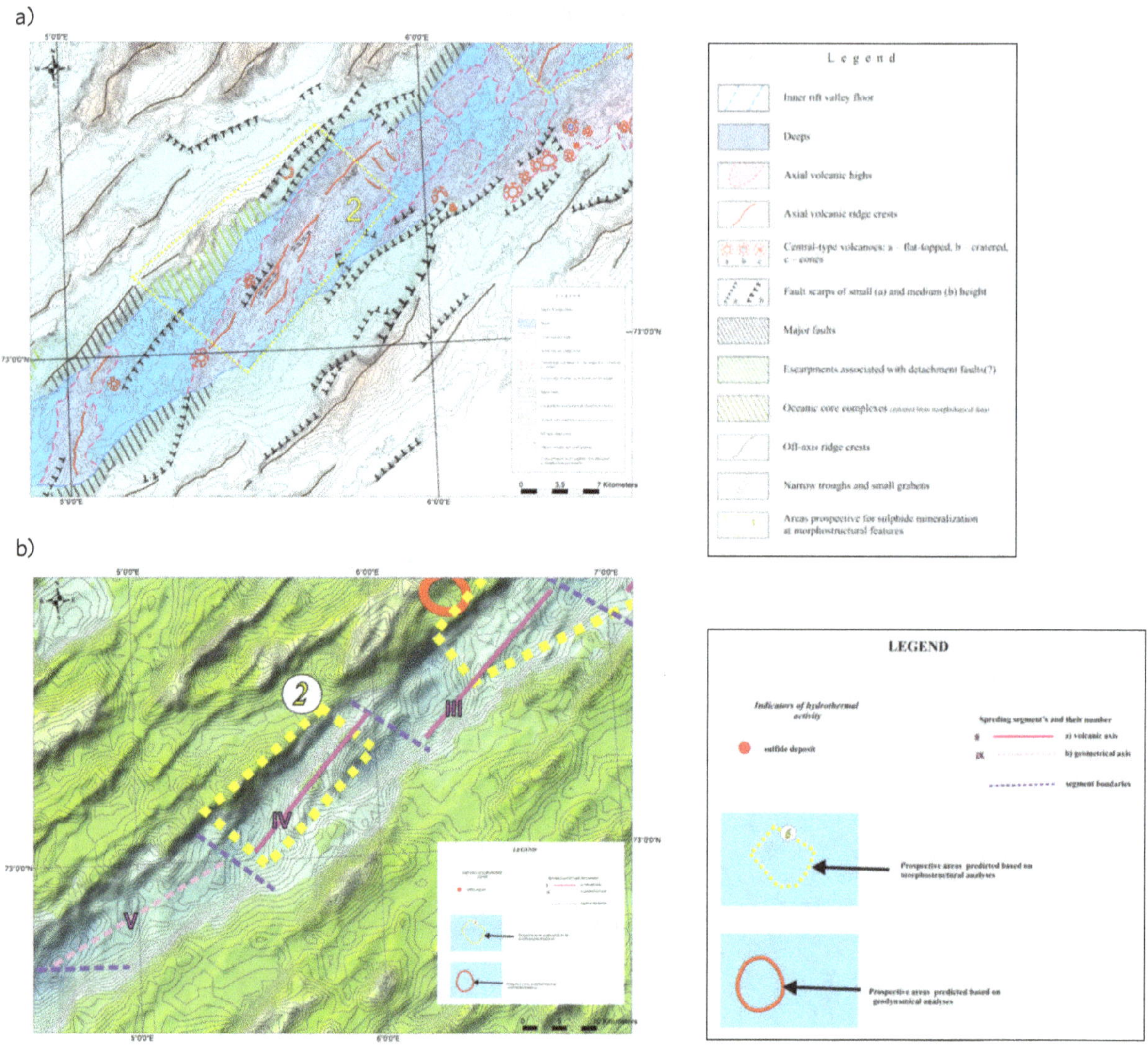

Figure 4-27 Mohns Ridge MS-M-area 2, southwest of interest area 4 defined by the geodynamical analysis. Figure a shows the morphostructural elements, whereas figure b shows the prospective areas identified by the geodynamic analysis, outlined in red. Click here and here for larger image.

This tract is solely defined from the morphostructural indicators. No significant geodynamic indicators are apparent.

4.2.8 Mohns – permissive tract M3 and MS-M-area 3

MS-M-area 3 lies in the central part of segment VI. See Figure 4–28. Its special feature is a very large and apparently rather old, massive volcano-tectonic ridge overlapping the eastern flank of the rift valley, and

74

linking with the structures of the western edge by a well-expressed transverse terrace. The promises for venting in this area are associated with the slopes of the western flank which, however, do not show many of the morphostructural indicators deemed necessary for hydrothermal manifestations. In the axial ridge, according to the morphostructural indicators and the lack of geodynamical indications of active venting, a relict ore occurrence may be expected.

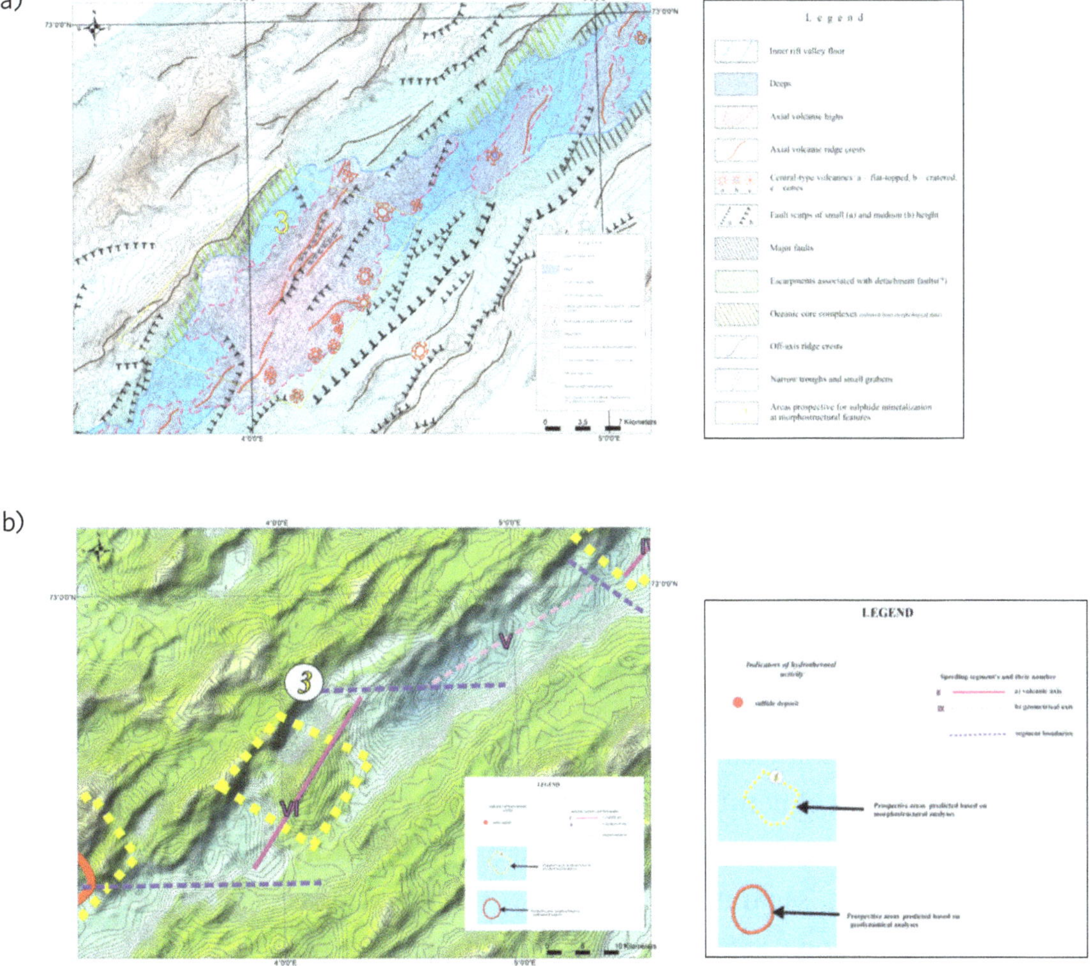

Figure 4-28 Mohns Ridge MS-M-area 3, northeast of interest area 5 defined by the geodynamical analysis. Figure a shows the morphostructural elements whereas figure b shows MS-M-area 3. No significant geodynamical indications are present. Click here and here for larger image.

As with tract M2, this tract is solely defined from the morphostructural indicators. No significant geodynamic indicators are present.

4.2.9 Mohns – permissive tract M4 and MS-M-area 4

The hydrothermal manifestation Copper Hill is in this area at 72°32′ N (Pedersen et al., 2010b). Copper Hill is located in the discharged zone, and lies near the highest gradient tension structures (black outline in Figure 4–29). See also Figure 4–30. Apart from the area connected to Copper Hill, there are two other areas with the same conditions that could be regarded as promising; indicated in Figure 4–29 by red outlines.

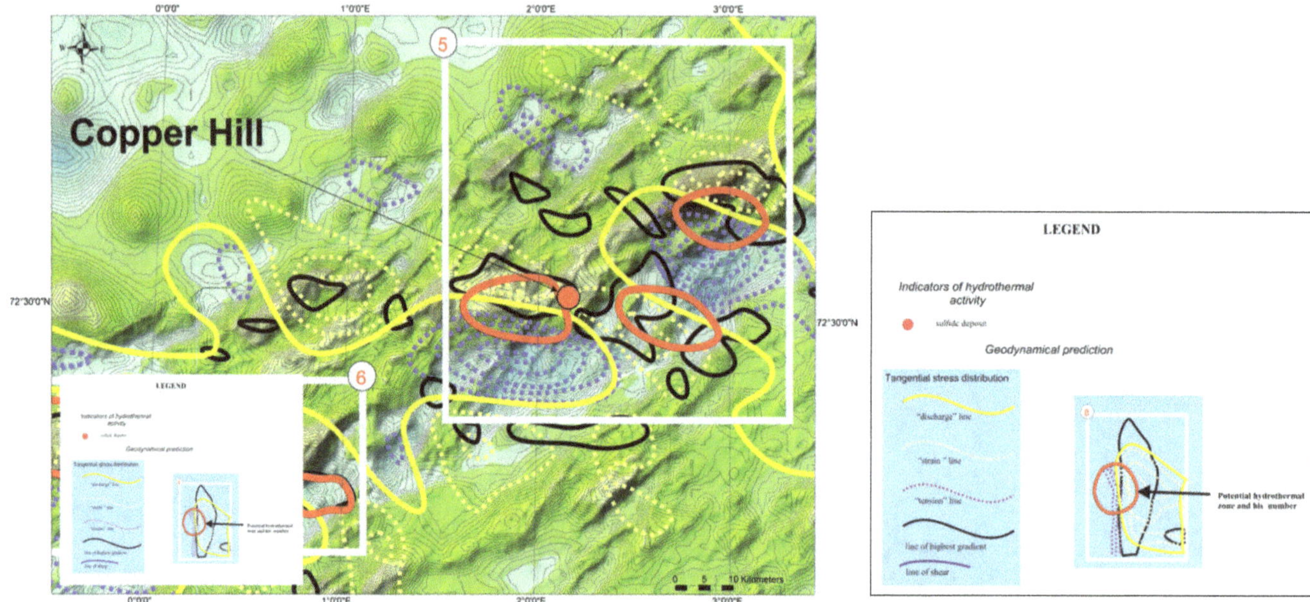

Figure 4-29 Mohns Ridge prospectivity of interest area 5, defined by the geodynamical analysis. Click here for larger image.

MS-M-area 4 (see Figure 4–30) fully covers segment VII and the adjacent zones of non-transform offsets of the rift valley, including the rather remote structures of its western flank. A funnel-like hole (crater?) in the depression, separating two (relict?) volcanic ridges is very noticeable in the axial zone. See Figure 4–30a. In the absence of an expressed volcanic rise, the presence of a crater may be indicative of a very shallow magma chamber (Vogt, 1974). The major prospects of this area are connected with the multiple, morphologically-attractive structures of the western flank, controlled by large faults.

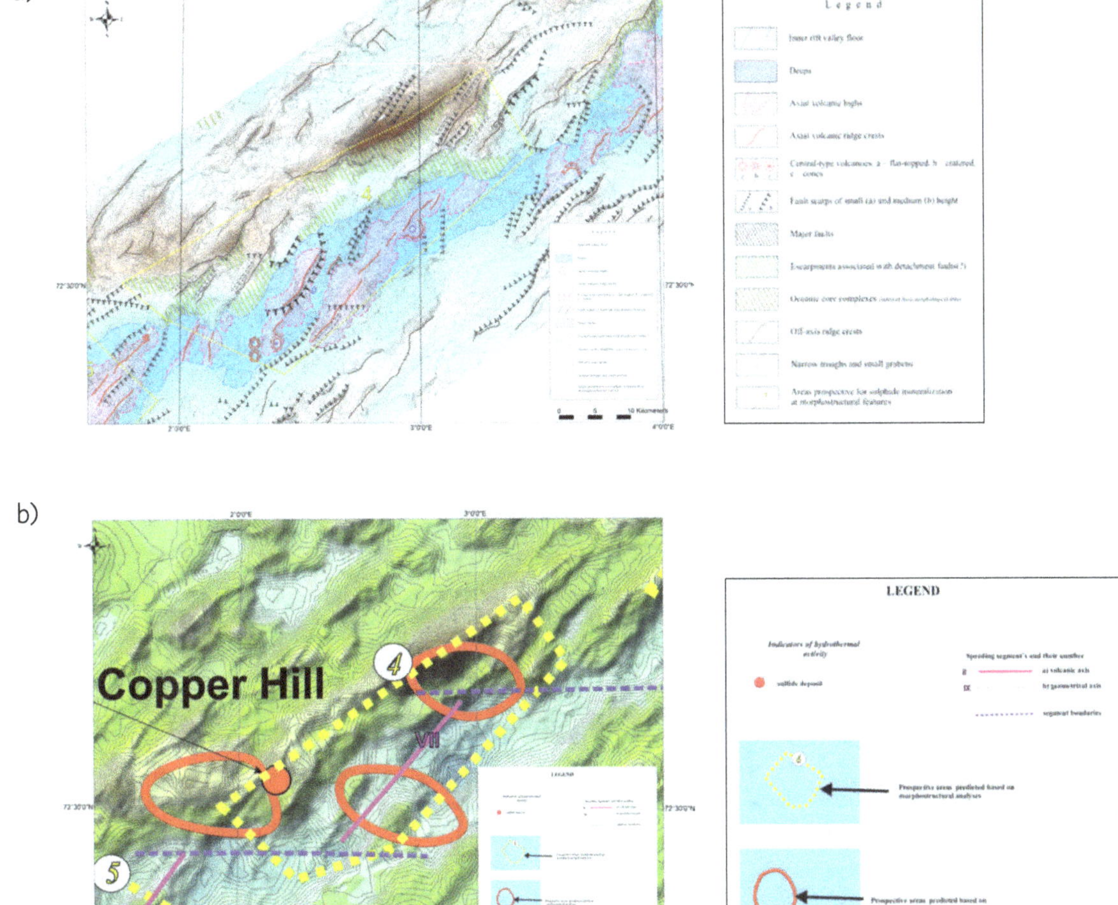

Figure 4–30 Mohns Ridge MS-M-area 4 (permissive tract 4), in area 5 defined by the geodynamical analysis. Figure a shows the morphostructural elements, whereas figure b shows the prospective areas identified by the geodynamic analysis, outlined in red. Click here and here for larger image.

4.2.10 Mohns – permissive tract M5 and MS-M-area 5

MS-M-area 5 is located in the central part of segment VIII, with many features similar to the MS-M-area 3 (segment VI). It is characterised by a well-defined structure with corrugations trending parallel to the spreading direction and occurring on a broad and smooth surface (see Figure 4–31a and b) of the western wall of the rift valley, making it comparable to a typical oceanic core complex.

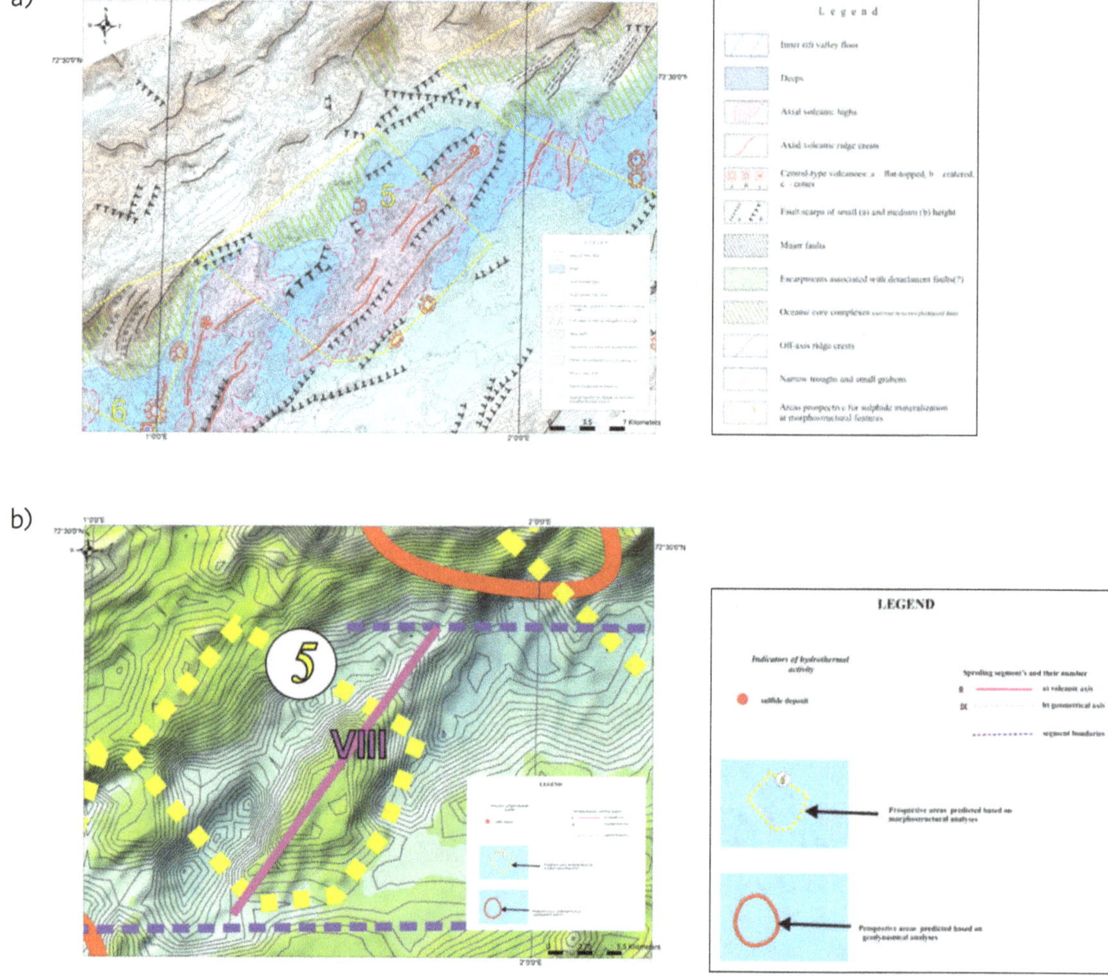

Figure 4-31 Mohns Ridge permissive tract M5, between interest area 5 and 6 defined by the geodynamical analysis. Figure a shows the morphostructural elements, whereas figure b shows the prospective areas identified by the geodynamic analysis, outlined in red. Corrugated structure interpreted in the north-western part of the tract. Click here and here or larger image.

No significant geodynamical anomalies are apparent in this tract, and it is solely defined based on the morphological analysis.

4.2.11 Mohns – permissive tract M6 to M9 and MS-M-area 6-9

Although no hydrothermal activity has been detected yet in this region, the results of the geodynamical analysis suggest that the area at the western flank of the ridge may be regarded as promising (the larger area within the red outline in Figure 4–32).

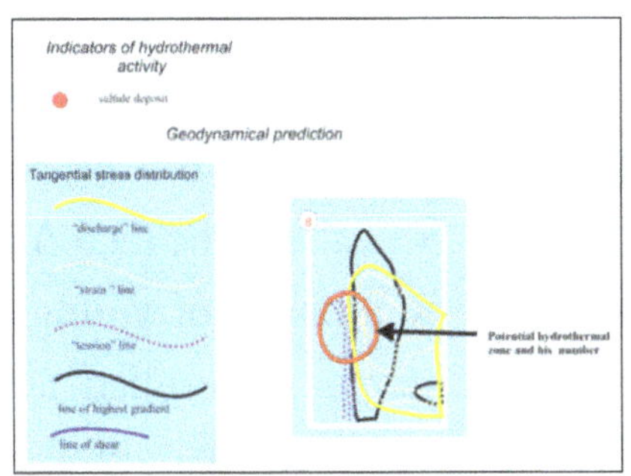

Figure 4-32 Mohns Ridge prospectivity of interest area 6, defined by the geodynamical analysis. Click here for larger image.

Connected to interest area 6 there are four promising permissive tracts; permissive tracts 6, 7, 8 and 9. See Figure 4–33. These tracts are, in the following, described separately.

a)

b)

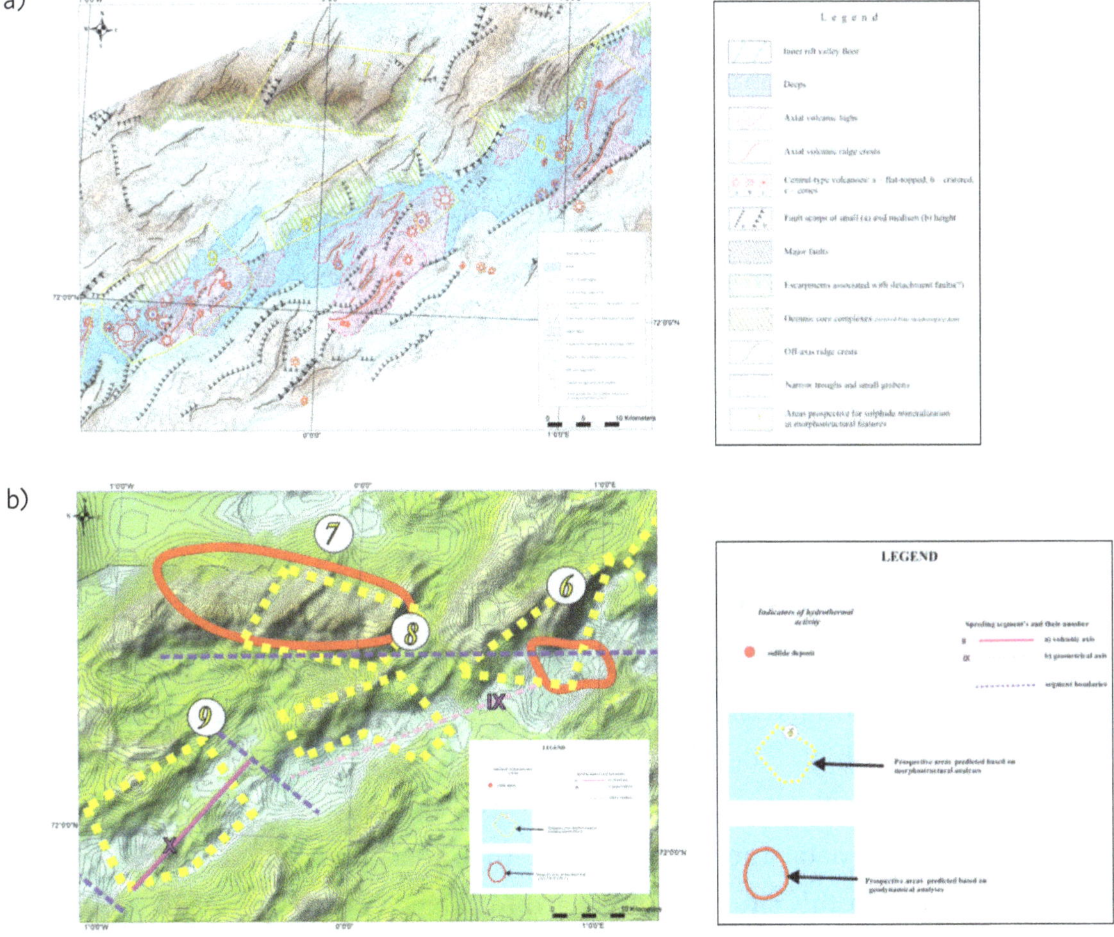

Figure 4-33 Mohns Ridge permissive tract M6 to M9, connected to interest area 6 by the geodynamical analysis. Figure a shows the morphostructural elements, whereas figure b shows the prospective areas identified by the geodynamic analysis, outlined in red. Click here and here for larger image.

79

4.2.12 Mohns – permissive tract M6 and MS-M-area 6

MS-M-area 6 is associated with a zone of the large lateral offset of the rift valley. The tectonic fault at the toe of a height is traceable southwards by a chain of young central-type volcanoes, which suggests its recent magma-controlled activity. Often, under such circumstances, certain extremely favourable conditions for high-temperature hydrothermal circulation appear in the area of the flank structures. In this case, sulphide deposits may occur in the graben-like depressions of the upper part of the elevation.

4.2.13 Mohns – permissive tract M7 and MS-M-area 7

MS-M-area 7 includes several relict highs of volcanic origin, similar to the above, formed at the earlier stages of the rift-valley development. The large transverse fault at their base seems to be still active. It is postulated that it is highly probable that relict sulphide deposits will be found here.

4.2.14 Mohns – permissive tract M8 and MS-M-area 8

MS-M-area 8 lies mainly on the western edge of segment IX of the rift valley. In the axial part of this segment, volcanic activity is rather low, apart from several large central-type volcanoes. This does not exclude any venting in the areas of detachment faults at the proximal flanks of the valley.

4.2.15 Mohns – permissive tract M9 and MS-M-area 9

MS-M-area 9 is located within the boundaries of segment X, which is very interesting from the morphological point of view, featuring an abnormally-high abundance of large flat-top central-type volcanoes on the valley floor and at the base of its western wall. According to the bathymetry data, the north part of this area represents an OCC-like structure. In this situation, it is hard to predict what type of hydrothermal circulation will prevail – the 'axial' or the 'flank' type. In general though, the area seems promising.

4.2.16 Mohns – permissive tract M10 and MS-M-area 10

No hydrothermal activity has been detected yet in this region. However, connected to area of interest 7, from the geodynamic analysis, there are three promising permissive tracts; numbers 10, 11 and 12. The line of discharge here coincides with pronounced tension structures (black outlines in Figure 4–34). The different permissive tracts are shown in Figure 4–36.

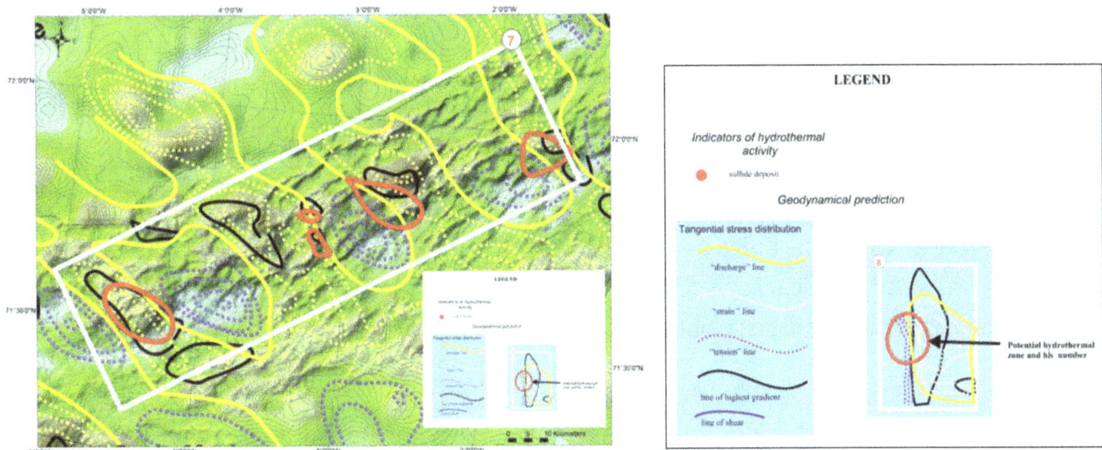

Figure 4-34 Mohns Ridge prospectivity of interest area 7, defined by the geodynamical analysis. Click here for larger image.

Permissive tract 10, like the preceding one, has many favourable morphostructural features, both in the axial zone and the western flank. Morphologically, its northers part looks like a large oceanic core complex and the area contains several volcanic rises. See Figure 4–35.

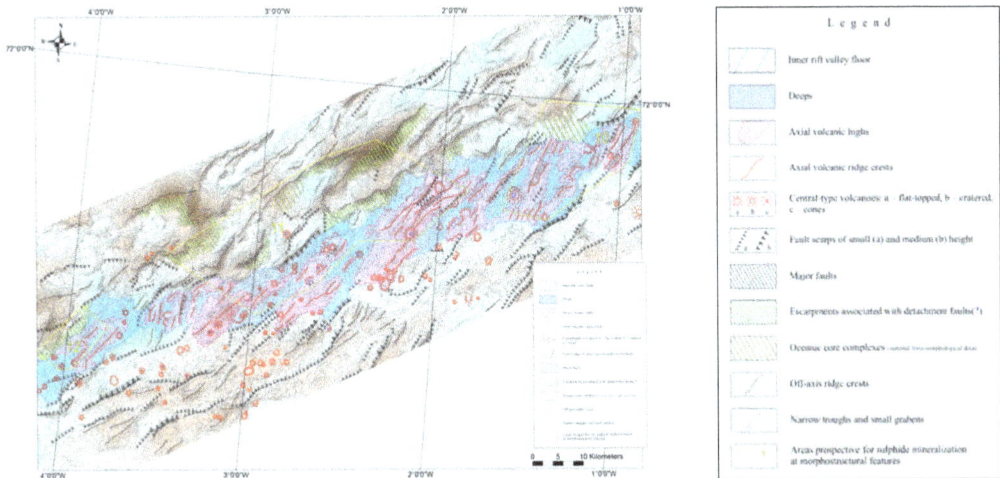

Figure 4-35 Mohns Ridge permissive tract M10 to M12, connected to interest area 7 by the geodynamical analysis. The figure shows the morphostructural elements. Click here for larger image.

81

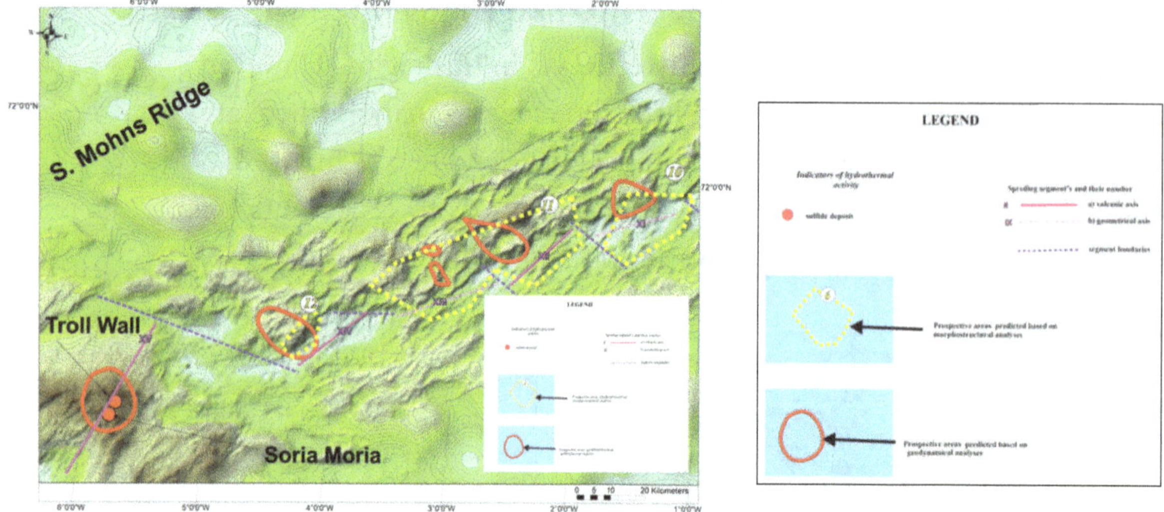

Figure 4–36 Mohns Ridge permissive tract M10 to M12, connected to interest area 7 by the geodynamical analysis. The figure shows the prospective areas identified by the geodynamic analysis, outlined in red. Click here for larger image.

4.2.17 Mohns – permissive tract M11 and MS-M-area 11

MS-M-area 11 combines multiple promising axial and flank structures of the two segments (XII and XIII) of the valley, including rather remote ones. See Figure 4–35 and Figure 4–36. It seems that its flank structures deserve special attention, since they are controlled by detachment faults demonstrating a variety of potentially ore-bearing morphostructural settings.

4.2.18 Mohns – permissive tract M12 and MS-M-area 12

MS-M-area 12 encompasses the western wall of the rift valley, defined by a detachment fault with steep inclination and a significant fault throw. See Figure 4–35 and Figure 4–36. At its base, a young volcanic zone is located and this fault may be interpreted as a magma-conducting channel. This situation suggests the high probability of hydrothermal fluid circulation in the zone of the flank tectonic dislocations, accompanied by the forma-tion of sulphide deposits in the apical part of the structure.

4.2.19 Mohns – permissive tract M13 and MS-M-area 13

One may see on the map that the location of the hydrothermal manifestations found earlier, Troll Wall (71°17' N) and Soria Moria (71°15' N) (Pedersen et al., 2010b), coincides with the intersection points of the discharge line (yellow) and a pronounced tension structure. This correlation confirms the results of the geodynamical analysis indicating possible locations of hydrothermal activity. See Figure 4–37. Inside interest area 8 there is a site, north of the Troll Wall site, where the different isolines cross or are close to each other. This site should be regarded as promising.

Figure 4–37 Mohns Ridge prospectivity of interest area 8, defined by the geodynamical analysis. Promising area north of Troll Wall, where the red, yellow and black lines cross. Click here for larger image.

Permissive Tract Characterisation

5.1 Deposit density

The well-mapped Troodos massif in Cyprus has been chosen as an analogue for the slow-spreading Mohns and Knipovich Ridges.

A difficulty in applying a density from ancient deposits to those currently at the MAR is to mimic as closely as possible how the permissive tracts of the MAR translate into the permissive area of the analogue. Singer et al. (2001) estimated the massive sulphide deposit density within the 1010 km^2 exposed permissive extrusive rocks from Troodos to be 20 deposits per 1000 km^2, with an ore tonnage larger than 20000 tons that all fit the Cyprus volcanogenic massive sulphide grade and tonnage model (Singer, 1986). This represents the necessary consistency between the grade–tonnage model and the density model that Singer (2014) underlines is critical for the estimated numbers to be meaningful. All of the 20 deposits are hosted within pillow lavas that can be assumed to represent the depositional environment of the slow-spreading MAR that forms the permissible assessment areas along the Mohns and Knipovich Ridges. Although Singer et al. (2001) state that it is unlikely that any undiscovered Cyprus volcanogenic massive sulphide deposits are exposed in extrusive rocks, since the exposures are very well-explored, continued drilling programs may locate buried deposits. Up to eight areas with high mineralisation potential have been outlined for future sub-surface exploration (Jowitt et al., 2012). Additional bias may also come from any pre-erosional amounts of sulphides that have been lost to erosion before reaching the present landform which, for the Iberian pyrite belt, is estimated to be 10–15% (Barriga et al., 1997). Mosier et al. (2007) state that effects of erosion cannot be demonstrated by lower deposit densities over time, but he does not discuss the effects of changes during the evolution towards the present landform. The volcanogenic massive sulphide database (Mosier et al., 2009) quotes that both the Agrokipia and Limni deposits have had some their massive zones of sulphides eroded away.

To accommodate possible biases, the deposit density of 0.02 deposits/km² for the extrusives of Troodos has, in this study, been modelled to correspond to the 20 percentile of the lognormal number of accumulation distribution (Table 1). The green dots in Figure 5–1 display these percentiles of the distribution on the plotted density of deposits per 100000 km² versus the permissive area in km² (based on data in Singer et al., 2010). The Troodos density is also included in the plot. Our choice of accumulation density distribution (Table 1) for the Mohns and Knipovich permissive areas spans the whole population of volcanogenic massive sulphide deposit densities for permissive areas of approximately the same size as that of the Cyprus extrusives.

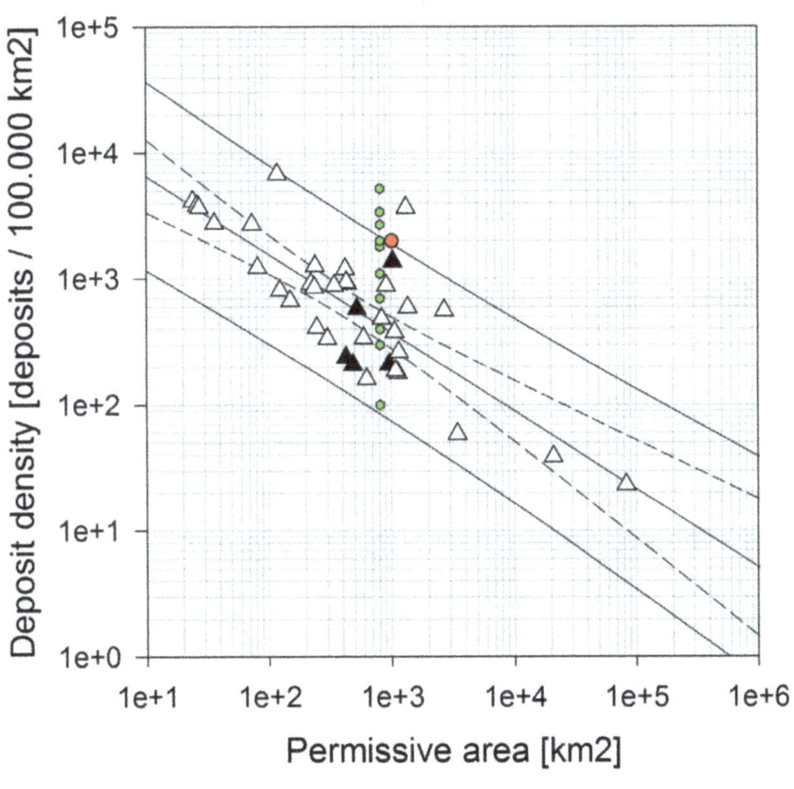

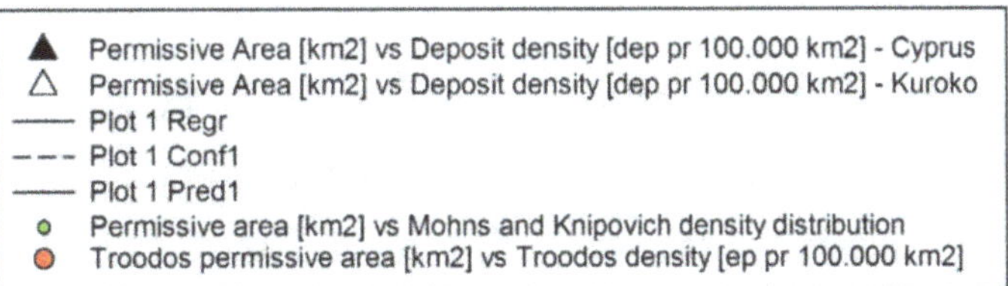

Figure 5–1 Plot of the density of deposits per 100.000 km² versus the permissive area in km². Percentiles of the SMS number of accumulations distribution in Table 1 is added to the plot with green dots. Figure based on data published in Singer et al. (2010). Troodos permissive area from Singer et al. (2001). Click here for larger image.

85

Assuming a lognormal distribution, the P20 and minimum and maximum values that span the population of volcanogenic massive sulphide deposit densities, the percentiles from the lognormal distribution can be derived. See Table 1.

Table 1 SMS density and aggregate number of accumulations inside the permissive tracts.

	Mode	Mean	Stdv	F100	F95	F90	F75	F50	F25	F20	F10	F5	F0
SMS density # acc per km²	0.007	0.014	10	0.001	0.003	0.004	0.007	0.011	0.018	0.020	0.027	0.034	0.052
# SMS acc	80	155	112	1	32	45	75	125	205	228	308	382	594

Given the density distribution, there is a 95% chance within the total 11389 km² of permissive area on the Mohns and Knipovich Ridges that there are 32 mineralised accumulations or more (percentiles can be seen in Table 1). The most likely number of accumulations, the mode, is 80 and the median is 125. The mean is as high as 155 due to the skew distribution, which has a coefficient of variation (CV) equal to 72%. This variability is of the same order of magnitude as the number of deposits distribution for mafic deposits, CV = 83%, in the recent assessment of the totality of SMS resources at mid-ocean ridges (MOR) by Singer (2014) and reflects the great uncertainty that accompanies assessments in frontier exploration areas.

The play methodology used for the assessment takes into account that only some of the hydrothermal activity indicators that will be investigated have the potential to lead to an SMS accumulation equal to, or larger than, the minimum size of 1700 metric tons of ore (50 metric tons of Cu metal) and a footprint equal to, or larger than, 0.04 km².

The number of hydrothermal activity indicators within each sub-play that could identify potential vent fields with surface expressions equal to, or larger than, 0.04 km², that could lead to SMS accumulations equal to, or larger than, 1700 metric tons, has been estimated to be proportional to the estimated number of accumulations and the size of the favourability zones, outlined by the morphological and geodynamical analysis. Due to the possibility of finding deposits well off-axis, like the TAG deposit connected to apparently long-lived boundary faults and small pillow volcanoes, some of the deposits could fall outside the stippled high favourability areas (outlined in Chapter 4).

The number of accumulations shown in Table 2 has been computed from the inferred, as well as the postulated, number of sites that could manifest hydrothermal activity within each of the 17 sub-plays, and subsequently aggregated to a total for the whole of the Knipovich and Mohns Ridges. The mean number has been estimated to be proportional to the deposit density and the size of the favourable areas outlined in Chapter 4. The expected total number of inferred and postulated hydrothermal manifestations in Table 2 has been calculated by dividing the number of accumulations by the probability 0.428 for having sulphides, given the presence of hydrothermal manifestations. The probability is calculated from the ratio between the number hydrothermal manifestations that show signs of sulphides and the total of hydrothermal manifestations along mid ocean ridges (Hannington et al., 2004 and 2013a) and the inferred and postulated manifestations is therefore the sum of manifestations with and without sulphides.

Table 2 Estimated number of accumulations and hydrothermal manifestations related to the permissive tracts (sub-plays).

Area	Permissive area [km²]	Area fraction of total area	Expected number of accumulations	Expected number of inferred and postulated hydrothermal manifestations
MS-K-area-1	3100	0.272	42.2	98.6
MS-K-area-2	2375	0.209	32.3	75.5
MS-K-area-3	375	0.033	5.1	11.9
MS-K-area-4	500	0.044	6.8	15.9
MS-M-area-1	1150	0.101	15.7	36.6
MS-M-area-2	200	0.018	2.7	6.4
MS-M-area-3	225	0.020	3.1	7.2
MS-M-area-4	713	0.063	9.7	22.7
MS-M-area-5	288	0.025	3.9	9.2
MS-M-area-6	150	0.013	2.0	4.8
MS-M-area-7	200	0.018	2.7	6.4
MS-M-area-8	138	0.012	1.9	4.4
MS-M-area-9	275	0.024	3.7	8.7
MS-M-area-10	325	0.029	4.4	10.3
MS-M-area-11	975	0.086	13.3	31.0
MS-M-area-12	100	0.009	1.4	3.2
MS-M-area-13	300	0.026	4.1	9.5
Sum	11389		155	362.1

Table 3 Play areas and distribution of number of accumulations. Total area 11,389 km².

ID	Area permissive tract [km²]	Inverse cumulative distribution of the potential number of accumulations										
		Mode	Mean	Std. dev.	F100	F95	F90	F75	F50	F25	F10	F5
MS-K-area-1	3100	13.5	42.5	43.5	0.0	5.0	7.9	14.9	29.0	54.0	94.0	129.0
MS-K-area-2	2375	7.0	29.2	25.3	0.0	3.8	5.8	11.0	21.0	39.9	65.0	84.0
MS-K-area-3	375	2.0	5.2	4.7	0.0	0.0	1.0	2.0	4.0	7.0	11.1	14.5
MS-K-area-4	500	2.0	6.6	6.5	0.0	0.0	1.0	2.0	5.0	9.0	15.0	19.6
MS-M-area-1	1150	4.0	16.6	14.4	0.0	1.9	3.0	6.4	13.0	22.0	35.6	45.7
MS-M-area-2	200	0.0	2.7	2.8	0.0	0.0	0.0	1.0	2.0	4.0	6.3	8.0
MS-M-area-3	225	0.0	3.1	3.2	0.0	0.0	0.0	1.0	2.0	4.0	7.0	9.8
MS-M-area-4	713	3.0	10.4	10.6	0.0	1.0	2.0	4.0	7.0	13.0	22.8	31.8
MS-M-area-5	288	1.0	4.0	4.2	0.0	0.0	0.0	1.0	3.0	6.0	10.0	12.0
MS-M-area-6	150	0.0	2.4	2.4	0.0	0.0	0.0	1.0	2.0	3.0	6.0	7.0
MS-M-area-7	200	0.0	2.7	2.8	0.0	0.0	0.0	1.0	2.0	4.0	6.3	8.0
MS-M-area-8	138	0.0	2.4	2.4	0.0	0.0	0.0	1.0	2.0	3.0	6.0	7.0
MS-M-area-9	275	1.0	4.0	4.2	0.0	0.0	0.0	1.0	3.0	6.0	10.0	12.0
MS-M-area-10	325	1.0	4.3	4.7	0.0	0.0	0.0	1.0	3.0	6.0	10.1	14.1
MS-M-area-11	975	4.0	13.6	12.1	0.0	1.0	2.0	5.0	10.0	18.2	28.7	36.7
MS-M-area-12	100	0.0	1.3	1.5	0.0	0.0	0.0	0.0	1.0	2.0	3.0	4.1
MS-M-area-13	300	1.0	4.0	4.2	0.0	0.0	0.0	1.0	3.0	6.0	10.0	12.0
Sum	11389		155									

A reliability check on these numbers can be made by looking at the growth in the number of discovered active vents over the last 45 years. The discovered number of active vent fields on the MOR (64000 km) grows each year (Figure 5–2), and can be predicted to reach more than 1000 over the next 100 years.

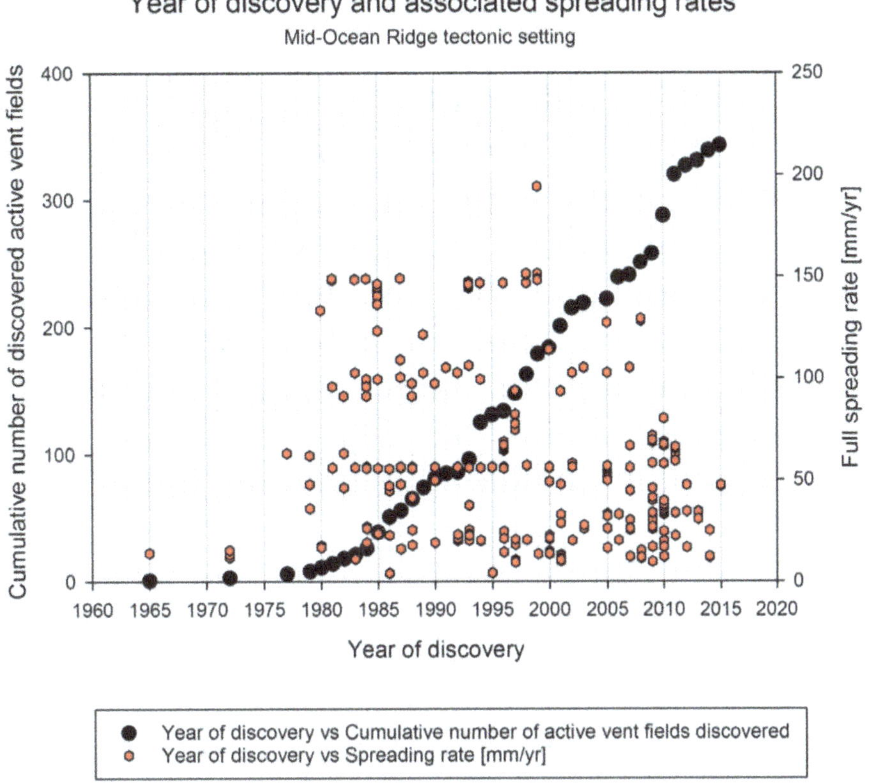

Figure 5-2 Evolution in time of the number of active fields on the MOR, based on data from Beaulieu et al. (2015). Click here for larger image.

If the length of the Knipovich Ridge is 460 km and Mohns Ridge is 570 km, then the 1030 km of ridge could, accordingly, be expected to ultimately have 16 active vent fields (being larger than the minimum used here). Inactive SMS accumulations, like 13030'N Semyenov, without hydrothermal manifestations are difficult to find. Postulating that, in addition to the surveyed and inferred vent fields, about an equal number of undiscovered vent fields might exist as concealed vent fields, with less bathymetric expression than the ones inferred from the morphostructural and geodynamic analysis of the bathymetric data. The 95% probability of having 32 or more SMS accumulations, as shown in Table 1, should therefore be regarded as a conservative projection for the ultimate number of both active and inactive undiscovered SMS accumulations on the Mohns and Knipovich Ridges, in view of the short SMS exploration period and low intensity of exploration until now.

89

Hannington et al. (2010) refer to estimations that all conclude that there are roughly 1000 active vent sites along the mid-ocean ridges. Furthermore, Hannington et al. (2010) conclude that the expected total number of what they term 'significant SMS occurrences along the neovolcanic zones of the world's oceans' is around 900, with a P90 of 500 and a P10 of 5000. Using the number of active vents at MOR settings in Beaulieu (2015) as an indicator of the distribution between the number of vents in different tectonic settings, around 500 out of these 900 could be found in MOR settings. This corresponds well with the minimum estimated value of deposits in neovolcanic zones presented in Hannington (2013a). Given 64000 km of mid-ocean ridge, this correlates to 8 active significant SMS occurrences per 1000 km of neovolcanic zones along the MOR.

5.2 Probability

Quantifying the probability on a regional level involves estimating the probability of favourable regional conditions that render it possible for local accumulations of metals. On the basis of the information from the known vent fields (Pedersen et al., 2010b) and the favourable and confirmed regional settings, this is set to 0.9 for unconfirmed plays. Confirmed plays have a play probability of 1.

Given the three SMS probabilities in Chapter 3, the first of these at the local scale is estimated to be 1, based on the criteria used for the seventeen permissive tracts outlined by the morphological and geodynamical analyses. This means that it is regarded as certain that there are local manifestations of hydrothermal activity, representing accumulations that potentially are within the required tonnage and grades inside the subplays, and furthermore, that these will be found if the plays are thoroughly explored.

The second of the individual SMS probabilities is estimated to be 0.428, given that from 327 sites with manifestations of hydrothermal activity, 140 sites showed active venting or polymetallic sulphides (Hannington et al., 2013a).

The third of the individual SMS probabilities is set to 1, based on our grade-tonnage and deposit-density requirement (Singer et al., 2008) that all accumulations must satisfy our minimum tonnage of 1700 metric tons used as the 100th percentile in the number of accumulations distribution shown in Table 6.

The SMS's technical discovery probability, given a favourable play, is the product of the probabilities for the adequacy of the three conditions above. This represents the probability that exploration activities can prove the presence of an accumulation with potentially exploitable amounts of metals, given the presence of hydrothermal activity related to the accumulation of 1700 metric tons or more of polymetallic sulphides with equal or larger metal content corresponding to minimum grades.

5.3 Tonnage and grade distributions

5.3.1 Introduction

The size and grade distributions that were used for the assessment of SMS on the Mohns and Knipovich Ridges are comparable to those compiled from Global SMS deposits at mid-ocean ridges and deposits in arc and back-arc settings, but have been modelled with less variability due to the expected lower frequency of extreme grades on the Norwegian ridges. The expected coefficient of variability is more than 3 times smaller than that calculated from the global data (Hannington et al., 2011b). The 25th percentiles for all these distributions are nearly half of those of the global SMS deposits, except for Cu where they are of similar magnitude. Copper has been modelled assuming a lognormal distribution with a 25% probability of having a grade of 7 wt % or more. It is important to underline that the grades represent average values over the whole tonnage of the SMS accumulation and are, therefore, lower than spurious grab samples from the surface or from individual samples.

The choice of grade and tonnage models should also take into account the geochemical redistribution related to how the metals are remobilised and re-deposited between pristine formations on the current oceanic crust and known VMS deposits in ancient terrains (e.g. Melekestseva et al., 2013).

5.3.2 Tonnage distribution

Hannington et al. (2010) present tonnages for 62 of the best-mapped SMS deposits. Only deposits that have surface areas larger than 100 m² (~3,000 tonnes, n = 42) have been included. These 42 deposits represent 67% of the deposits included in Hannington et al. (2010). The intercept for the 50th percentile indicates a median deposit size of 70,000 metric tons. Hannington et al. (2010) also present cumulative size frequency distribution of SMS deposits at MOR. This includes the 10% largest deposits, like Middle Valley, Zenith-Victory, Puys des Folles (~10 x 10⁶ t), Semyenov (~9 x 10⁶ t), Krasnov (~3 x 10⁶ t), Sunrise (~3 x 10⁶ t), the TAG Mound (2.7 x 10⁶ t), and Solwara 1 (2.1 x 10⁶ t). The Atlantis II Deposit in the Red Sea is excluded.

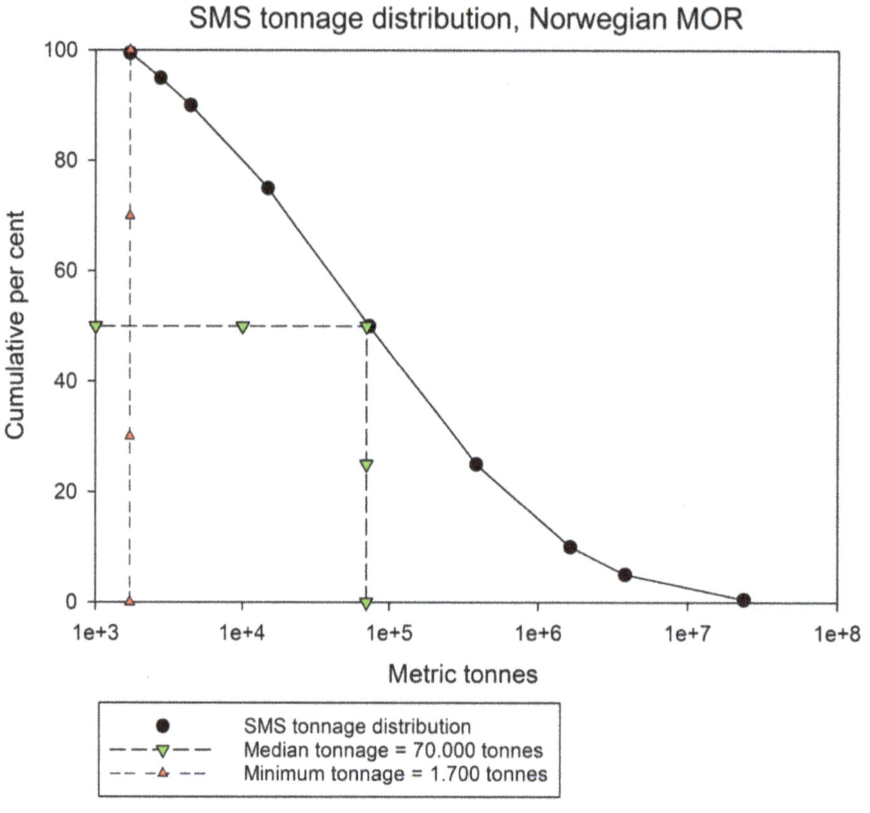

Figure 5-3 Cumulative size frequency distribution of sea floor massive sulphide (SMS) deposits used in this study. Click here for larger image.

Figure 5–3 and Table 4 show the size distribution used in this study.

Table 4 Ore-tonnage distribution

	Mode	Mean	Std. Dev.	F100	F95	F90	F75	F50	F25	F10	F5	F0
Ore tonnage [10³t]	2.0	855.4	3053.2	1.7	2.8	4.5	15	73	383	1643	3824	23856

It was decided to keep the same median size, of 70000 metric tonnes, as the distributions in Hannington et al. (2010). The minimum size is set to about 1700 metric tonnes, corresponding to a surface area of 40 m² of exposed sulphides. This is similar to the 90% percentile of the curve in Hannington et al. (2010). This choice is motivated by the requirement for having accumulations with a similar minimum contained metal in a SMS deposit on the Mohns and Knipovich Ridges as in the grade-tonnage models (Figure 5–4) in the recent assessment of the total MOR (Singer, 2014).

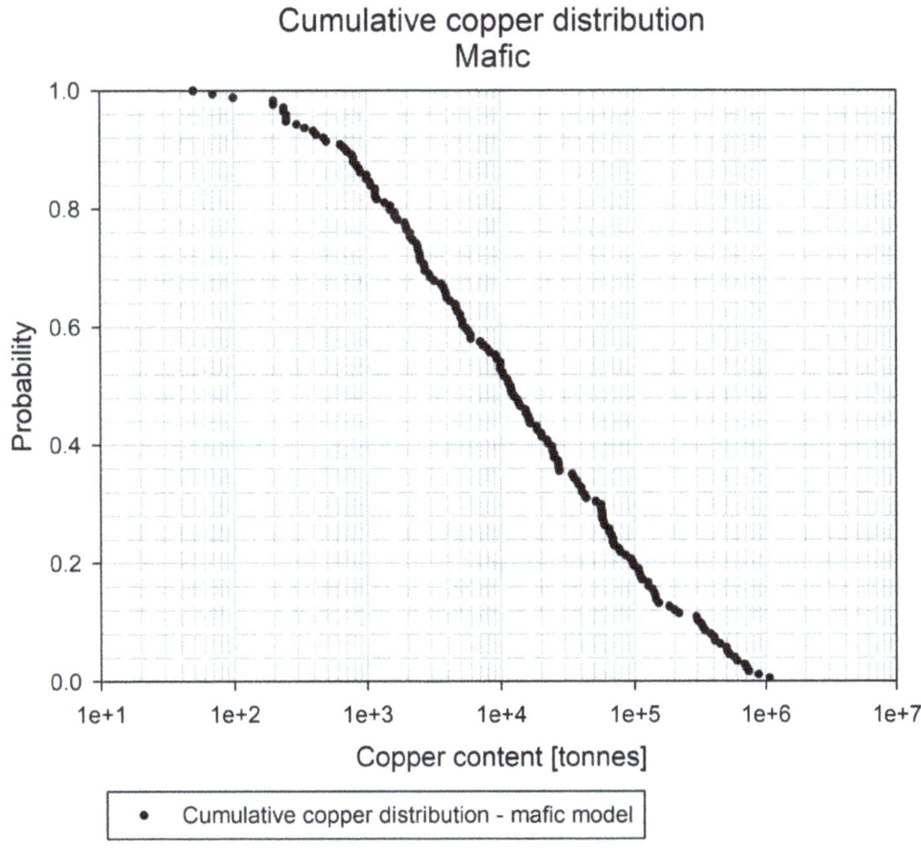

Figure 5-4 Metal distribution in mafic massive sulphide deposits. Data from Mosier et al. (2009). Click here for larger image.

5.4 Grades

The choice of grade distributions (Table 7) for SMS on the Mohns and Knipovich Ridges for Cu, Zn, Au and Ag deviates from the Global MOR distributions in Table 6 (Hannington et al., 2011a). The 18 deposits in our analogue area from the Troodos ophiolite have coefficients of variability (CV%) for Cu of 60%. The Mohns and Knipovich grades have coefficients of variability of less than 50%, except for gold with 121%. The grades in Table 6 all have coefficients of variability (CV%) larger than 80%, and a Cu/Zn ratio that is too much influenced by samples from intra-oceanic back-arc and arcs, as well as transitional arcs and continental margin arc environments, to represent valid distributions for our assessment.

Table 5 Average bulk compositions of 95 SMS deposits, 3,869 surface samples (from Hannington et al., 2013a).

Geological Setting	N	Cu [wt%]	Zn [wt%]	Pb [wt%]	Au [ppm]	Ag [ppm]
Mid-ocean ridges	2071	5.9	6.1	<0.1	1.6	89
Sedimented ridges	173	1.1	3.6	0.5	0.5	84
Intra-oceanic back-arc	668	3.9	16.4	0.9	6.6	210
Intra-oceanic arc	169	5.3	17.7	2.4	9.6	407
Transitional arc	728	6.4	14.8	2.0	12.2	692
Continental margin arc	60	3.1	20.3	10.0	2.3	953

Table 6 Percentiles for SMS grade distributions for SMS deposits, all tectonic settings (from Hannington et al., 2011a).

SMS (N=62)	75th Percentile	50th Percentile	25th Percentile	Mean	Std. dev	Std. dev/ Mean %
Cu metal [wt %]	1.70	4.30	7.00	5.58	4.61	83 %
Zn metal [wt %]	4.30	10.60	18.60	15	14.9	99 %
Au metal [ppm]	0.30	1.70	5.30	7.05	28.38	403 %
Ag metal [ppm]	37.00	107.00	234.00	209.8	353.7	169 %

Table 7 Percentiles for SMS grade Norwegian MAR

Metal grades	75th Percentile	50th Percentile	25th Percentile	Mean	Std. dev	Std. dev/ Mean %
Cu metal [wt %]	5.04	5.78	6.62	5.90	1.35	23 %
Zn metal [wt %]	4.88	5.84	6.94	6.08	1.82	30 %
Au metal [ppm]	0.61	1.06	1.82	1.58	1.91	121 %
Ag metal [ppm]	62.3	82.9	109	90.9	44.1	49 %

The grade distribution that was used for the assessment of SMS on the Mohns and Knipovich Ridges was modelled as shown in Table 8.

Table 8 Grade distributions for individual SMS accumulations.

Metal grades	Mode	Mean	Std. dev	F100	F95	F90	F75	F50	F25	F10	F5	F0
Cu metal [wt %]	5.45	5.90	1.35	3.00	3.93	4.33	5.04	5.78	6.62	7.56	8.25	10.90
Zn metal [wt %]	5.74	6.08	1.82	2.66	3.62	4.10	4.88	5.84	6.94	8.33	9.38	13.34
Au metal [ppm]	0.67	1.58	1.91	0.07	0.23	0.34	0.61	1.06	1.82	3.18	4.58	12.8
Ag metal [ppm]	76.7	90.9	44.1	19.2	37.1	46.5	62.3	82.9	109	143	171	284

The individual distributions are shown in Figures 5–5 and 5–6.

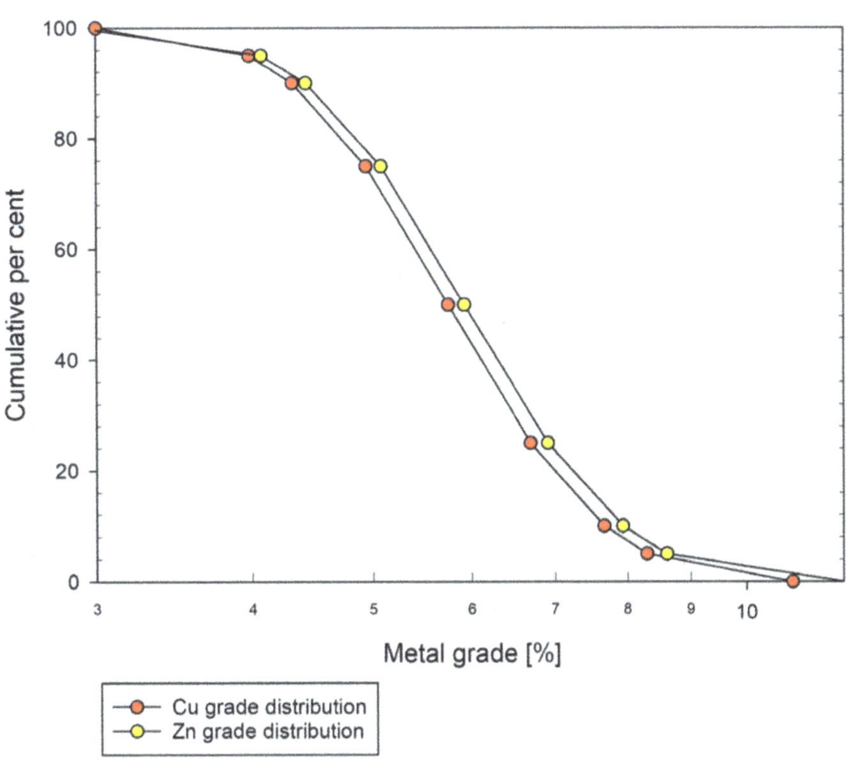

Figure 5–5 Copper and zinc grades. Mean grades given in Table 8. Click here for larger image.

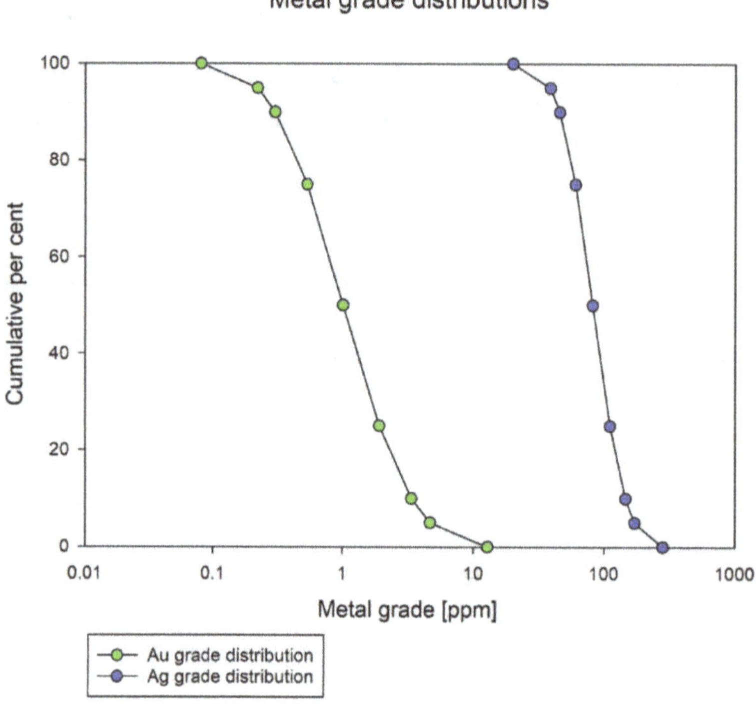

Figure 5-6 Ag and Au grades. Mean grades given in Table 8. Click here for larger image.

Given the grade and tonnage distributions, one can calculate the distribution of contained metric tons of metal in a single undiscovered SMS accumulation. See Table 9.

The metal distributions in the SMS deposit model from the Mohns and Knipovich Ridges (Table 9) are similar to the mafic models used by Singer (2014), especially for high-end percentiles, but deviates below the upper quartile (See Figure 5-4). If we compare the medians in Table 9 with Figure 5-4, which are based on Singer (2014), we note that the models used in this study represent a compound model dominated by the mafic model for the higher tonnages, and the felsic-bimodal model for the lower tonnages.

Table 9 Distribution of contained metric tonnes of metal in a single undiscovered SMS accumulation

Resource Type [Units]	Mode	Mean	Std. dev.	F95	F75	F50	F25	F10	F5	F1	F0
Cu metal [tonnes]	146	50684	187408	159	859	4209	22318	95677	221013	956006	1445386
Accumulation size											
Au metal [tonnes]	0,0	1,4	8,0	0,0	0,0	0,1	0,4	2,0	5,1	24,5	39,6
Accumulation size											
Zn metal [tonnes]	152	52122	191568	164	871	4303	22843	98204	229640	979139	1401190
Accumulation size											
Ag metal [tonnes]	0,2	76,7	300,1	0,2	1,2	5,9	32	143	335	1421	2099
Accumulation size											

In view of the great uncertainties regarding pristine grade and tonnages of VMS accumulations, this study chose deposit densities from the Cyprus VMS deposits. For grades, the project chose the mean values according to the average values from mid-ocean ridges, and the maximum values according to the observed tonnages and grades in the Russian MAR exploration areas (Cherkashov et al., 2010, 2013). It must, however, be noted that there are only a few drilled deposits. Sample material comes primarily from surface samples. The appropriateness of the grade distributions used here has been assessed by comparing the total amount of metal potential with the total metal potential calculated by Hannington (2013b), Cathles (2013) and Singer (2014). See Chapter 7.

Results

6.1 Assumptions

The probabilistic methodology was applied to each of the 17 sub-play assessments to produce metal estimates in the form of probability distributions. The sub-plays are assumed to belong to the same geological play, i.e. the MAR, even if local ridge segments are individually different. To take into account the limited dependency that can be inferred between sub-plays, aggregation (summation of the potential in the 17 sub-plays) was conducted with dependency factors.

The Knipovich Ridge shows similar architecture to the north-eastern Mohns Ridge but spreading along the Knipovich is up to 60° oblique, and the axial valley ridges (AVR) are more widely spaced, typically 100 km, compared with 30–50 km at the central Mohns Ridge (Rona et al., 2010). Due to the narrower spacing of the AVRs, more diversity can be expected along the Mohns Ridge than along the Knipovich Ridge. A dependency factor of 0.45 is therefore used for Mohns play aggregation, compared to a higher dependency factor of 0.6 for the aggregation of the Knipovich plays. These dependency factors regulate the dependencies between the sub-plays within each of the two ridges.

If one or more of the Mohns Ridge sub-plays will produce economically-recoverable amounts of Cu, Zn, Au and Ag, then the likelihood is high for having favourable mineralisation conditions also in one or more of the Knipovich sub-plays. A dependency factor of 0.9 is therefore used when the resources of the Mohns and Knipovich Ridges are combined to represent the total expected metal resources from the Norwegian part of the MAR. This dependency factor regulates the dependency between the two ridges.

In the assessment of in-situ gross monetary value, a gram to ounce (oz) ratio of 28.35 has been used to accommodate the gold price quotes (USD / oz).

6.2 Metal resource potential

Each of the 17 sub-play probability distributions for Cu, Zn, Au and Ag were aggregated using a Monte Carlo simulation (GeoX, 2014), to produce resource estimates for the entire Mohns and Knipovich Ridges (Table 10). For each commodity, the following distributions are calculated:

- Aggregated number of accumulations
- Accumulation size distribution
- Conditional aggregate potential, given that the assessment area has at least one sub-play with a confirmed SMS
- The unconditional aggregate potential.

The conditional aggregated potential estimate presupposes that at least one of the 17 sub-plays are confirmed, whereas the unconditional aggregated potential does not make this assumption. The unconditional potential is therefore lower than the conditional potential because the latter is multiplied by the probability that the plays are confirmed. The metal distributions are all skewed to the left, generally with a standard deviation equal to or substantially larger than the mean (for aggregate potential).

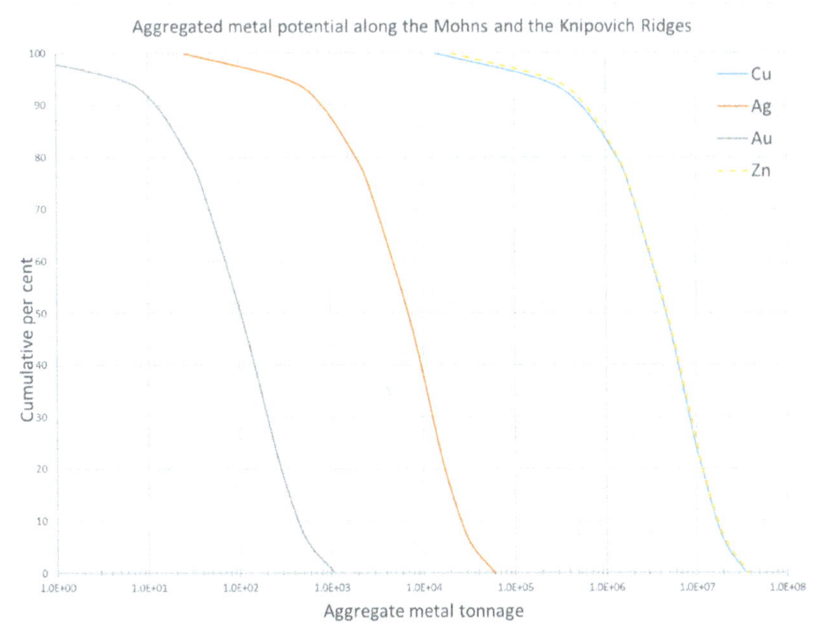

Figure 6-1 Aggregated metal potential along the Mohns and Knipovich Ridges. Click here for larger image.

Table 10 shows the conditional and unconditional aggregated potential. The mean of the skewed distributions is approximately 7 million metric tons of copper and zinc, 175 metric tons of gold and about 10,000 metric tons of silver. Figure 6–1 shows the metal distributions, illustrating that there is a 5% probability that the metal potential along the 1030 kilometres of studied ridge is about 21 million metric tons of copper, 22 million metric tons of zinc, just under 600 metric tons of gold and 34000 metric tons of silver or more. The uncertainties and the skewness are large, illustrated by the large standard deviations and the large difference between the minimum and maximum values (min. / max. values illustrated only in Figure 6–1). For silver, copper and zinc, the ratio between the expected value, the mean, and the standard deviation is slightly above 1 or 100%. For gold, this ratio is, as expected, significantly higher than one. The copper and zinc distributions in Figure 6–1 are overlapping. The zinc distribution is illustrated with a dotted yellow line. Table 10 also show the percentiles P95, P75, P50 (the median), P25 and P5. For copper, these indicate that there is a 95% chance that the aggregated copper tonnage along both the Mohns and Knipovich Ridges is larger than ~200 thousand metric tons. For zinc, there is a 5% chance that the tonnage is ~23 million metric tons or more. For gold there is a 95% chance that he aggregated tonnage is larger than 5 tonnes and a 5% chance that it is larger than ~600 tonnes. For silver the corresponding tonnages are 340 and 34000 tonnes for 95% and 5% respectively. Given the skewness of the distributions, which is also apparent in Figure 6–1 (remembering that these distributions are plotted on a logarithmic scale) the 'mode' is the most likely value and significantly smaller than the mean.

We observe in Table 10 that the conditional and unconditional aggregate potentials are nearly equal, indicating that there is a negligible risk that none of the 17 sub-plays will confirm at least one SMS accumulation. The assessment results will therefore only show the distributions for the conditional aggregate commodity potential that was used as a basis for calculating the total gross metal value for each commodity.

Table 10 Aggregated accumulation number, size and in-situ metal resources on the Mohns and Knipovich Ridges

Resource Type [Units]	Mode	Mean	Std. dev.	F95	F75	F50	F25	F5
Cu metal [metric tons]								
Number of Cu accumulations	92	161	124	30	74	126	210	416
Accumulation size Cu	0	45834	137948	0	1680	4585	21791	224533
Conditional aggregate potential Cu	14558	6884938	7033961	195906	1752433	4613919	9614568	21678550
Uncond, aggregate potential Cu	14558	6882185	7033901	194216	1749347	4611112	9611228	21675930
Au metal [metric tons]								
Number of Au accumulations	72	157	117	28	74	126	211	385
Accumulation size Au	0	1	3	0	0	0	0	5
Conditional aggregate potential Au	0	175	201	5	38	103	238	597
Uncond, aggregate potential Au	0	175	201	5	38	103	238	596
Zn metal [metric tons]								
Number of Zn accumulations	68	161	121	29	74	129	213	402
Accumulation size Zn	0	51913	160575	6	1667	4446	21340	257048
Conditional aggregate potential Zn	36855	7095203	7338270	258783	1776559	4703893	9945984	22746144
Uncond, aggregate potential Zn	36855	7092365	7338174	256571	1773835	4700878	9943090	22743314
Ag metal [metric tons]								
Number of Ag accumulations	73	161	122	29	74	127	210	399
Accumulation size Ag	0	68	207	0	2	6	33	338
Conditional aggregate potential Ag	25	10524	11237	339	2528	6821	14527	34074
Uncond, aggregate potential Ag	25	10514	11236	332	2519	6813	14517	34064

6.3 Metal prices

Metals are traded on stock exchanges worldwide, and their prices are controlled by varying supply and demand mechanisms and are highly volatile on local and regional scales. It is, however, outside the scope of this book to do a robust and in-depth forecast and sensitivity analysis of future metal prices. The basis for the metal price assumptions are prices reported by Thomson Reuters / SNL Metals and Mining. See Figure 6–2.

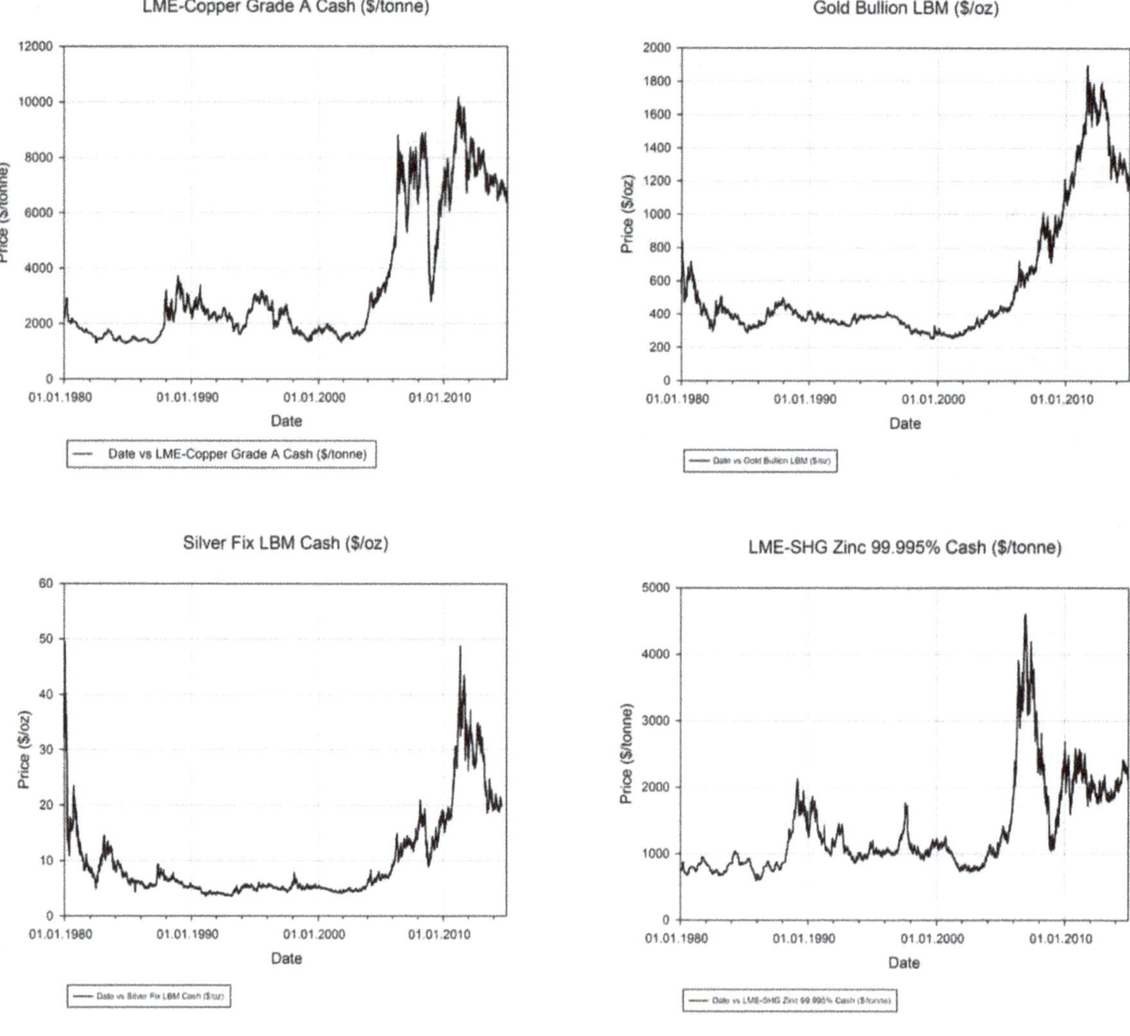

Figure 6–2 Metal prices from January 1980 until recently. Used as a basis for deciding the commodity prices used in the valuation. Source: Thomson Reuters, SNL Metals & Mining, an offering of S&P Global Market Intelligence. Click here, here, here and here for larger image.

The following rounded values close to the prices at the time of the assessment (15/12-2016) have been assumed in the calculation of the gross value:

- Au=1200$/oz
- Ag=20$/oz
- Cu=7000$/metric ton
- Zn=2000$/metric ton

6.4 Gross value of yet-to-find metal resources along the Knipovich and Mohns Ridges

The aggregated gross values (in-situ metal value) for copper, zinc, gold and silver are given in Table 11.

The metal gross values are given by multiplying the 'Conditional aggregate potential' given in Table 10 with the commodity prices in Table 11.

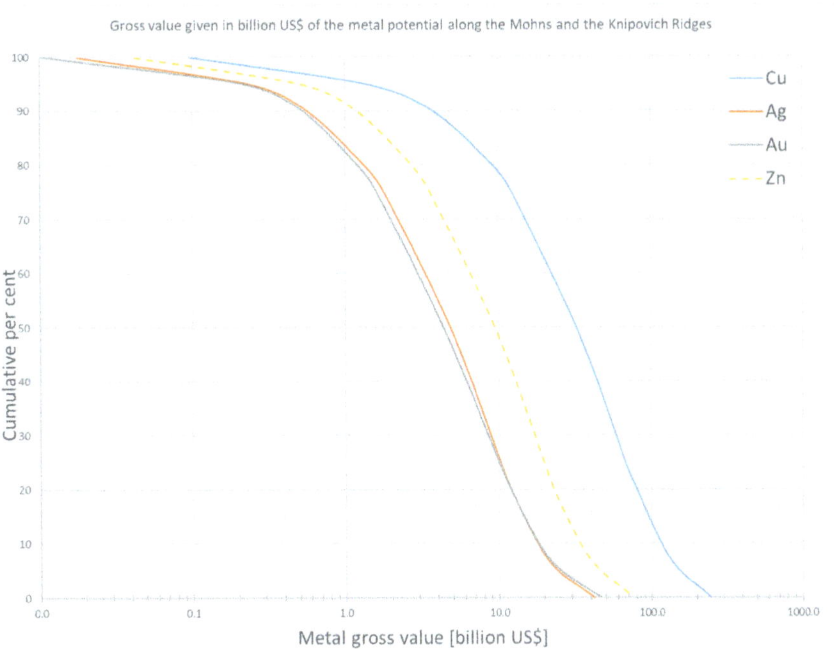

Figure 6-3 Aggregated gross value of copper, zinc, gold and silver along the Mohns and Knipovich Ridges. Click here for larger image.

Figure 6–3 and Table 11 show the gross value distributions. The figure illustrates that given the commodity prices above, the *in-situ* value potential along the 1030 kilometres of ridge is between US\$ 0.1 billion and US\$ 246 billion for copper (blue graph in Figure 6–3), between US\$ 0.01 billion and US\$ 46 billion for gold (grey graph in Figure 6–3), between US\$ 0.04 billion and US\$ 75 billion for zinc (dashed yellow graph in Figure 6–3) and between US\$ 0.02 billion and US\$ 41 billion for silver (orange graph in Figure 6–3). This gives a total upside potential of about 400 billion US\$. Clearly, these numbers are associated with a high degree of uncertainty, given both the uncertainty in commodity prices, the deposit density, the deposit tonnage and the deposit grades. It is also apparent in Figure 6–3 that copper is the major financial driver.

Table 11 show the percentiles P95, P75, P50 (the median), P25 and P5 for the Aggregated gross values of the commodities. For copper, these quantify a 95% chance (P95) that the gross copper value (*in-situ* value) along both the Mohns and Knipovich Ridges is larger than 1.4 billion US\$. For zinc, there is a 5% chance (P5) that the gross value is 45.5 billion US\$ or larger. For gold and silver there is a 5% chance that the *in-situ* value is larger than 25 billion US\$ and 24 billion US\$ respectively. The distributions are highly skewed, with consistently slightly higher standard deviation than mean and a much lower mode than mean, emphasising the uncertainty.

The gross values are compared to the Norwegian Oil Fund. This fund had a market value of 8,300 billion Norwegian kroner (NOK) by November 5, 2018. The mean gross value of the total metal resources along the Mohns and the Knipovich Ridges is thereby about 8% of the Norwegian Oil Fund.

Table 11 Aggregated gross value Mohns and Knipovich Ridges

Resource Type and gross value [Units]	US$ / metric ton	Mode	Mean	Std. dev.	F95	F75	F50	F25	F5
Cu metal gross value [Billion US$]									
Conditional aggregate potential	7000	0.10	48.2	49.2	1.4	12.3	32.3	67.3	151.7
Au metal gross value [Billion US$]									
Conditional aggregate potential	42328754	0.009	7.4	8.5	0.2	1.6	4.4	10.1	25.2
Zn metal gross value [Billion US$]									
Conditional aggregate potential	2000	0.07	14.2	14.7	0.5	3.6	9.4	19.9	45.5
Ag metal gross value [Billion US$]									
Conditional aggregate potential	705479	0.018	7.4	7.9	0.2	1.8	4.8	10.2	24.0
Total gross metal value [10^9 US$]		0.2	77.2	80.3	2.3	19.2	50.8	107.5	246.5
Norwegian Oil Fund 5/11-18 [NOK and US$]. Exchange rate 8.37	8.3E+12	9.9E+11							
% of Norwegian Oil Fund 4/1-17		0.02 %	7.80 %	8.12 %	0.02 %	1.94 %	5.14 %	10.86 %	24.91 %

Validation of Aggregated Play Resources

Despite excellent reviews (e.g. Pedersen et al., 2010b; Hannington, 2005–2013), it remains difficult to distinguish the fundamentals from the interesting details regarding the ultimate potential for sea-floor massive sulphide (SMS) deposits. Several influential publications have addressed the issue of estimating the resource base of sea-floor massive sulphide (SMS) deposits (Cathles, 2011a, 2011b; Hannington et al., 2011b; Hannington 2011, 2013b; Singer, 2014; Juliani & Ellefmo, 2018a). From a cursory evaluation of these and related articles, it may appear that some authors are optimists, whereas others are pessimists. Thus Cathles (op. cit.) estimates that with an accumulation efficiency of 3% (3% of the metals in the hydrothermal fluids is precipitated as sulphides), the amount of copper in sea-floor mineral deposits, 106 billion metric tonnes that would last for more than 6000 years at the current rate of extraction, whereas Hannington and co-authors found the amount of copper and zinc in the more easily-accessible (i.e. surface deposits) of the neovolcanic zones of the global oceans to be just 30 million metric tonnes. The large discrepancy in the two estimates is not only a matter of optimism versus pessimism, but rather reflects the fact that two different reservoirs have been estimated. Cathles evaluates the total metal retained in mineral deposits in the oceanic crust (exposed plus buried), whereas Hannington focuses his attention on surface deposits in the more easily-accessible neovolcanic zones of the global oceans. The estimate by Cathles differs in two main aspects:

- the whole area of the global oceans is included; and
- the implicated mineralisation is not only on, but also under the sea floor.

From an ore deposit perspective, questions are mainly reduced to:

- What fraction of the metal carried by the venting fluid is actually accumulated on the sea floor (what is the accumulation efficiency)?
- To what extent are the accumulated metal sulphides concentrated spatially (how are the sulphides swept into piles)?

Cathles' starting point is the fact that the heat introduced into the crust at the ridge equals that lost by 350°C venting. This means that the thickness of the crust alone determines the mass of 350°C seawater that can circulate, on average, through each square metre of sea floor. He then combines this with the observed chemistry (from MORs in the Pacific) of the high-temperature axial vent fluids, and calculates that 0.67 kg/m^2 of copper metal could be on each square metre of sea floor.

The crustal thickness along Mohns Ridge is well below the global average for typical oceanic crust, and shows a high variability with a mean thickness of around 4 km (Klingelhöfer et al., 2000). With 2D seismic imaging of the sub-seafloor and a given density model, Krysiński et al. (2013) documented that the oceanic crust formed at the Knipovich Ridge is between 6 and 8 km thick.

The total deposition of Cu in the neovolcanic zone (including known sub-sea-floor stockwork mineralisation and near field sediments) is, according to Hannington (2013b), ~1.74 kg/m^2, which comprises: 0.44 kg in near field sediments, 0.34 kg in SMS deposits and 0.96 kg in stockworks. Using the tonnage factor of 3 indicated above, this corresponds to 5.19 kg/m^2, which comprises: 1.31 kg in near field sediments, 1.01 kg in SMS deposits and 2.87 kg in stockworks, for the total Cu deposited in the neovolcanic zone along an ultra-slow ridge. This corresponds to about 9% of the total high-temperature axial flow, assuming a minimum flux to the ridges of ~58.2kg Cu/m^2. Following Hannington's (2013b) statements about the balance and the adjustments to MAR conditions, the difference between the minimum flux and the total deposition of 5.19 kg/m^2 in the neovolcanic zone of 53 kg/m^2 Cu must be deposited in the crust as sub-sea-floor mineralisation associated with diffuse axial flow.

The discrepancy in the two estimation approaches of Cu metal in SMS deposits/m² applied to the Mohns–Knipovich Ridges may, according to Barriga et al. (2013), not be a matter of optimism versus pessimism, but rather reflects the fact that Cathles' approach, leading to 1.88 kg Cu/m², evaluates the total metal retained in mineral deposits in the oceanic crust (exposed plus concealed). On the other hand, Hannington's approach leads to 3.88 kg Cu/m² for SMS and sub-sea-floor alteration and mineralisation (1.01 kg Cu/m² + 2.87 kg Cu/m²). Computed exclusively for the 'significant massive sulphide accumulation' *on* the sea floor, in the easily-accessible neovolcanic zone, this value of 1.01 kg Cu/m² gives 10.4 million metric tons of copper, which corresponds to the 22% percentiles for Cu in the aggregated resources in Table 10, and a smaller upside on the assessed resources than that calculated using Cathles' approach.

Singer (2014) makes a probabilistic estimate of the Cu, Zn, Pb, Au and Ag tonnages inside a 4 km narrow neovolcanic zone around the ridge, with a total area of 256,000 km². He uses the three-part assessment framework and onshore analogues to model the contained metal and deposit density (number of deposits per km²). Estimations are thereby computed with much smaller deposit densities and only pertain to the 4 km wide zone, with resulting Cu resources that represent a clear downside which corresponds closely to the 95% percentiles for Cu in the aggregated resources in Table 10.

Figure 7–1 shows a comparison between the potential resource of copper in percentiles, obtained by this study, and the mean values obtained by Cathles (2013), Hannington (2013b) and Singer (2014). One may notice that the percentiles defined in this study span the sample space defined by the three authors, yielding a degree of further confidence in our estimates.

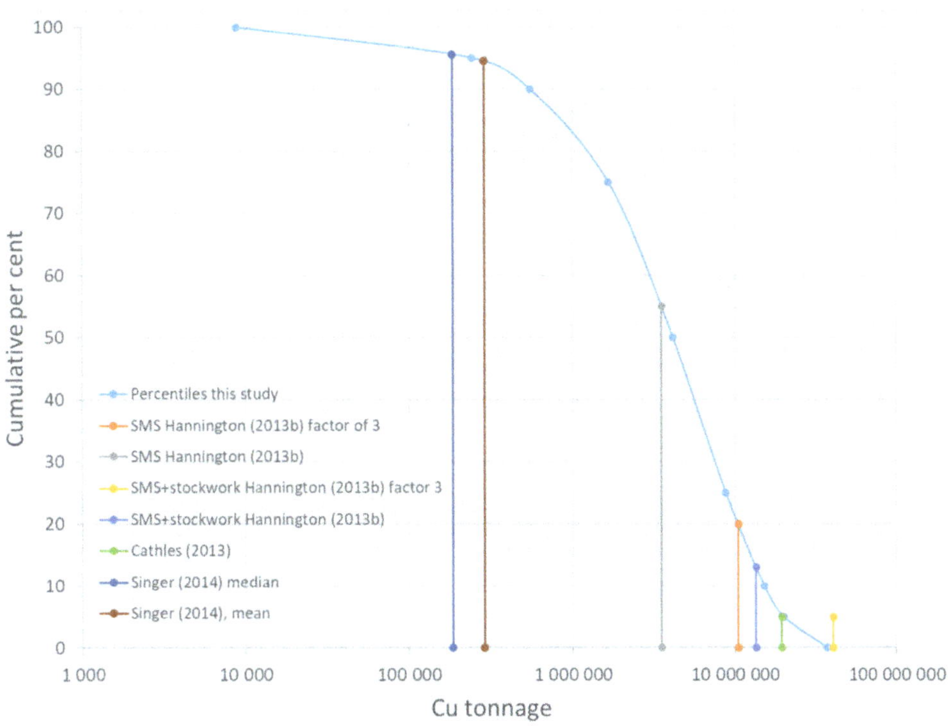

Figure 7–1 Validation of Cu tonnage obtained in the mineral resource copper potential estimates inside the Norwegian Extended continental shelf. Click here for larger image.

Testing the Hypothesis

8.1 The MarMine Project

To reduce the uncertainty of the presented estimates of the mineral resource potential along the extended Norwegian continental shelf, more exploration and sampling is needed.

NTNU therefore applied for a grant from the Norwegian Research Council, together with 13 Norwegian companies, in 2015. The main goals of the MarMine Project were to test exploration technologies and to collect mineralised material to seek to improve the accuracy of the estimate of undiscovered mineral resources on the extended Norwegian continental shelf, focusing on seafloor massive sulphide (SMS) deposits along the Arctic Mid-Ocean Ridge (AMOR), and to evaluate exploitation technologies for potential future mining activities. An integrated part of this is to test to see whether an exploration strategy based on a morphostructural analysis of the ocean floor bathymetric data can be used to find undiscovered accumulations of sulphide mineralisation. In the summer of 2016, the MarMine Project carried out a three-week research cruise to the AMOR (Ludvigsen et al., 2016) with the multipurpose subsea vessel, Polar King (see Figure 8–1).

The main purpose of the research cruise was to collect samples for further analysis and in the course of its duration, several hundreds of kilos of collapsed chimney fragments were collected from the Loki's Castle mounds. Most of these were collected from the surface, but the project also used a Remotely Operated Vehicle (ROV) mounted drilling system to obtain sub-surface samples. One of the interesting research questions is whether the mineralisation differs at different depths. However, core drilling on the Loki's Castle mounds was difficult due to the mixture of unconsolidated sediments and chimney fragments (Ludvigsen et al., 2017; Søreide et al., 2017). One of the Loki's Castle chimneys can be seen in Figure 8–2.

Figure 8-1 M/V Polar King multipurpose subsea vessel with the Hugin AUV surfacing with new data from the seafloor. Photo: NTNU MarMine; no reuse without rights-holder permission. Click here for larger image.

Figure 8-2 One of the chimneys on the Loki's Castle site. Hydrothermal fluid can be seen flowing out of the chimney. Photo: NTNU MarMine; no reuse without rights-holder permission. Click here for larger image.

To investigate the appropriateness of Autonomous Underwater Vehicles (AUV) as a platform to find hydrothermal manifestations, multivariate datasets were collected from the Loki's Castle site, the Mohns' Treasure area, and two exploration areas south-south-east of Mohns' Treasure. The benthic fauna at the presumed inactive site Mohns' Treasure was naturally distinctively different (low biomass, high biodiversity) from the active site at Loki's Castle (high biomass and low biodiversity). Sponges and crinoids on rocky substratum and stalked crinoids on sediments dominated. Echinoderms (small asteroids and holothurians) were also observed on sediment. An Axial Volcanic Ridge (AVR) was defined as exploration area 1 and 2. This exploration area definition was primarily done based on the morphostructural work presented herein identifying faults, graben structures and volcanic features. Based on high-resolution bathymetric maps, geological features were selected for ground-truthing using the ROV. A dramatic bathymetry, similar to alpine terrain, was identified but no new hydrothermal manifestations were discovered during these surveys.

8.2 Post-cruise results

In order to better understand the composition of SMS, it is necessary to directly investigate the material of which they are comprised. Physical samples are needed to make microscopically-thin sections to understand the mineralogy, and to make powders suitable for chemical analysis using X-ray and electron beam techniques. On terrestrial sites of geological interest, this is a relatively simple process, but at a depth of 2500 metres under water in the middle of the Arctic Ocean, physical sampling is more complicated. During the MarMine research cruise, AUVs and ROVs were sent to the ocean floor to investigate the sites, with the aim of collecting samples and data to return to the laboratory.

The metal-laden fluids escape back into the ocean from the crust via holes and cracks referred to as hydrothermal vents, much like when we see geysers or gas escaping from vents in hot water spring areas. As the hot, buoyant metal-enriched water is expelled from the hydrothermal vents, it forms tall sediment-laden plumes. Vent plumes dominated by Zn and Pb are known as 'white smokers', and those dominated by Cu and Fe as 'black smokers'. The minerals which precipitate out of the solution from these vents build up over time (like inside a kettle), literally forming chimneys for the vents. These chimneys are composed of the elements carried by the hydrothermal fluids, and therefore the minerals which form in these chimneys are likely to be representative of those which form in potentially-significant quantities at lower depths. This is clearly visible at Loki's Castle, where several chimneys are present in the two main mounds, and some of these chimneys are active black smokers.

The hydrothermal vent chimneys are prone to collapse periodically, due to internal fluid-pressure building up and subsequently breaking off fragments, earthquakes that destroy the structures completely, or simply because the smokers grow too large to support their own weight. These fragments of the vent material fall down the sides of the mounds, forming scree slopes and debris fields.

It is this material that was successfully collected by the ROV and eventually investigated by the project scientists, along with samples of the muddy sediments from the sea floor, and drill cores of basalt which represent the oceanic crust.

In total, more than 200 kg of loose boulders from the flanks of the mound and collapsed fragments from the chimneys were recovered. Due to the often elevated levels of Fe in the SMS material, the rocks have a strong rusty coating masking the internal features. Upon first cutting and coring the boulders, initial impressions were all related to the variability of the material. The rock is often highly porous, with the cavities lined with a variety of crystals. The texture is chaotic, with billowing laminations of rusty and white carbonate, sulphate and silicate minerals, grey clay-dominated areas, and fine-grained black zones. It is the black zones which typically contain the elements and sulphide minerals of interest: copper in chalcopyrite and isocubanite, zinc in sphalerite, and, occasionally, silver and gold. In some material, these phases (including pyrite) amalgamate into interlocking lenses, hence the term 'massive sulphide'. These can be seen in Figure 8-3, where coarser sulfide layers are hosted within a porous, black fine-grained groundmass (scale bar is 4 cm).

Figure 8-3 Metallic layers hosted within a porous, black fine-grained groundmass (scale bar is 4 cm). Click here for larger image.

Initial petrographical and geochemical investigations, based on the surface samples from Loki's Castle, indicate the potential for grades and tonnages comparable to those of economically-significant volcanogenic massive sulphide (VMS) deposits; ancient terrestrial versions of SMS deposits which have long been utilised as a major source of Cu, Zn and Pb from onshore mines.

The majority of the samples which display visible sulphide mineralisation typically contain 0.5wt% Cu and 1wt% Zn, but samples with as much as 2wt% Cu, and 7.4wt% Zn have been recorded. These grades are considerably lower than the average grades used in the resource potential estimation presented herein, but the amount of data these samples represent is not assessed to be great enough to change the defined grade distributions. In order to really lower the uncertainty, even more data must be collected and, in particular, along the 3rd dimension (i.e. sub-seafloor). Both Ag, Au and, surprisingly, some Pb are present in the collected material. Understanding the form of the Cu and Zn-bearing minerals is important in order to provide information necessary for subsequent mineral-processing considerations, ensuring efficient extraction. More information on mineralisation, content and potential exploitation technologies will become available as the MarMine Project continues.

8.3 Updating and improving the play definitions

During the MarMine research cruise, data from Loki's Castle, the greater Mohns' Treasure area, including the scarp edges, and two exploration areas on an axial volcanic ridge were collected. Collected data included high resolution sea floor bathymetry, sediment acoustic reflectivity (acoustic backscatter), salinity, temperature, pressure, turbidity, magnetism and methane content, using the Multi-Beam EchoSounder (MBES), CTD, turbidity sensor, Self-Compensating Magnetometer (SCM) and CH_4-sensor. This data will undergo detailed studies to improve the understanding, descriptions and definitions of the plays included in the play analysis. In the future, this will reduce uncertainty because it will be feasible to develop more specific play analyses, e.g. an AVR, a sediment hosted, and an oceanic core complex play analysis.

8.4 Governmental exploration cruises to Mohns Ridge

Early in 2018, the Norwegian Petroleum Directorate (NPD) asked for qualifying documents to participate in a request for tenders in relation to the *'Acquisition of Seabed Massive Sulfide Data in the Norwegian Sea'* (NPD, 2018a). After the tendering process was concluded, NPD engaged Swire Seabed AS and Ocean Floor Geophysics Inc. to map for potential sulphide mineralisations inside a predefined area, along central parts of Mohns Ridge (NPD, 2018b). The defined area overlaps significantly with permissive tract M3 and M4 defined in this study. See Chapter 4.2. NPD required spontaneous potential data (SP), multi-beam bathymetry (MBES), backscatter data, and grav/mag with high frequency seismic (HFS) and sub-bottom profiler (SBP) data as optional. Electromagnetic (EM) and magnetotelluric (MT) data were also defined as relevant. In addition, relevant geological samples from areas of special interest, sampled with gravity cores or grabs, were required (NPD, 2018a, and NPD, 2018b).

The area defined by NPD is interesting from a deposit perspective. It would be expected that the collection of high resolution bathymetric data would give local and pertinent details of the sea floor morphologies rich in textural information. The northern and southern AVRs may reveal fissured hummocky terrains with intermittent growth of lava domes and cones (several to hundreds of metres in diameter), or smooth lava flows where eruption rates are higher. In addition, larger quasi-circular volcanic structures (1 km in diameter, 100–200 m height) may disclose local fissuring and terrains that go from relatively smooth upper surface (flat-topped type) to steep-sided topographic depressions (crater type). Vertically-elongated shapes rising up from the sea floor may also dissect the hummocky terrains, i.e. eruptive fissures which consist of melt flows that are often in conjunction with other features, such as faults. The sediment thickness, which is expected to be low in these young and volcanically-active areas, increases with distance away from the active spreading centre as the crust gets older and less volcanically active. Combined with acoustic backscatter images, important details can be tied to sedimentary processes along all the multibeam bathymetry lines, which provide information on local mass movements or sediment-covered flat sea floor that is often found in arch-shaped depressions. If used, the sub-bottom geo-acoustic profiler may help decipher the underlying rugged topography, and differentiate areas of little sediment accumulation to those where sediment transport is higher, notably in structurally-controlled off-axis provinces. Off-axis normal faulting may expose valuable outcrops that preserve distinguishable mineralised horizons. Gravitational collapses and landslides may, however, participate greatly in covering smaller features, especially within tectonic terrains of large fault scarps.

The focus of the exploration cruise was not to find active systems, but rather inactive systems assumed to still contain mineralised material. The project commenced on 1st August 2018, involving the ROV support vessel Seabed Worker from Swire Seabed AS and a Kongsberg Hugin AUV with various sensors from Ocean Floor Geophysics Inc. After 25 days, the project was successfully completed finding new deposits, both associated with active and inactive systems (NPD, 2018c). It is stated that '*The NPD has identified a large area of sulphide minerals, that was previously unknown. The deposits could include important industrial metals such as copper, zinc, cobalt, nickel, vanadium, wolfram and silver.'* Furthermore, *'The new area of sulphide minerals that was discovered contains many such piles of gravel and collapsed black smokers, including a 26 metre high, non-active tower, in addition to some active systems.'* This confirms the interesting potential along the Mohns Ridge. Figure 8.4 shows the approximate location of the newly discovered sulphide occurrence in the central parts of the Mohns ridge.

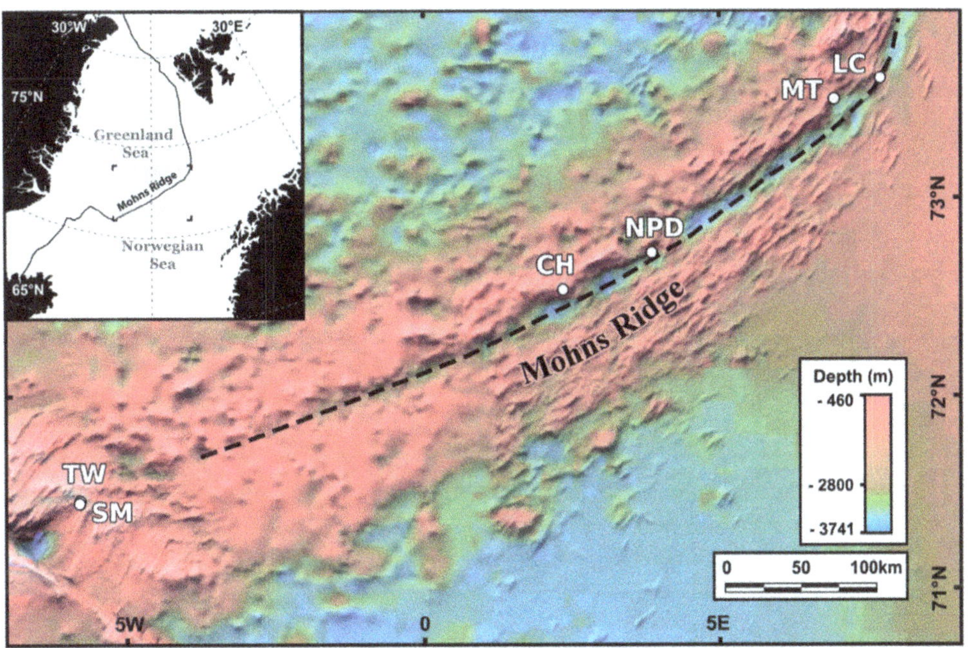

Figure 8-4 Location of the newly discovered sulphide occurrence ('NPD' in map) in the central parts of the Mohns Ridge. The newly discovered occurrence is plotted with the occurrences Loki's Castle (LC), the Mohns Treasure (MT), Copper Hill (CH), the Troll Wall (TW) and the Soria Moria (SM). Ægir vent field is not included. Source map: NTNU / Cyril Juliani. Source occurrence location: Norwegian Petroleum Directorate. Click here for larger image.

The location confirms play M3 (see section 5.2) which enables an update of the resource potential estimates already build, using the play analysis framework. In a new analysis, the sub-play M3 is then confirmed and will therefore be associated with a higher play probability, i.e. a lower regional scale risk. No grade data have been made available and collected sample material will in the coming years undergo detailed mineralogical, textural and chemical analysis. A reliable tonnage estimate requires drilling in combination with geophysics to understand the occurrence geometry and grade distribution. The necessary data for this is not yet available.

During another exploration cruise to Mohns Ridge in August-September 2019, NPD discovered additional 'new sulphide deposits' (NPD, 2019). Exact location and characteristics have not yet been published but the exploration area overlaps permissive tracts identified and described herein. Simultaneous deployment of more than one AUV enabled timely, cost-efficient data collection, and large amounts of high resolution data were collected. This is promising, an when more information from the cruise is made available, the present resource potential estimate can be updated.

Discussion and Concluding Remarks

9.1 Classification of the marine minerals discovered on the AMOR

Deposits can be defined both according to geological knowledge and according to the so-called modifying factors that include the mining method, processing route and environmental, social, economic, governmental, market and legal aspects. US Geological Survey (USGS, 1980) classified mineral resources into identified and undiscovered resources (Figure 9.1). Mineral resources that are identified are further separated into inferred, indicated and measured resources, dependent on the knowledge base. Undiscovered resources are postulated and can be, like identified resources, sub-divided dependent on the level of geological knowledge into hypothetical and speculative resources. Hypothetical resources are postulated resources in confirmed regions, whereas speculative resources are inside unconfirmed regions.

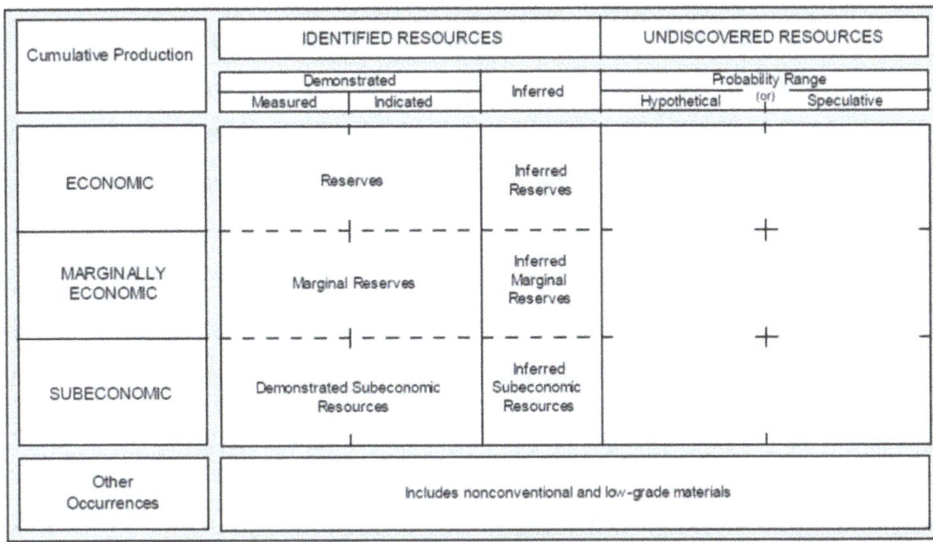

Figure 9–1 USGS classification scheme for mineral resources and reserves. Figure from USGS (2009). Click here for larger image.

Worldwide, there are a number of classification systems for mineral resources and reserves in use. The National Instrument 43-101 (NI, 2018) and the Australasian Joint Ore Reserves Committee Code (JORC Code; JORC, 2012) are two commonly-used standards. The JORC Code is aligned with the classification scheme of the USGS (1980), and classifies resources and reserves into the following sub-groups:

- An inferred mineral *resource* is a deposit which has been sampled sufficiently to make an estimate of its grade and tonnage at low confidence.
- An indicated mineral *resource* is a deposit where the continuity is not yet confirmed but in which tonnage and other deposit characteristics can be estimated with a reasonable level of confidence.
- A measured mineral *resource* is a sampled deposit with a high level of geological confidence and confirmed geological continuity.
- A probable ore *reserve* is the economically-mineable part of an indicated or a measured mineral resource, dependent on the uncertainty in the modifying factors.
- A proven ore *reserve* is the economically-mineable part of a measured mineral resource. The deposit has been proved mineable also in terms of the modifying factors.

JORC (2012) continues with a definition of an exploration target. This is defined as an '*estimate of the exploration potential of a mineral deposit in a defined geological setting where the statement or estimate, quoted as a range of tonnes and a range of grade (or quality), relates to mineralization for which there has been insufficient exploration to estimate a Mineral Resource*'.

The mineralisation that has been located on the AMOR could be termed as exploration targets (JORC, 2012), whereas the mineral resource assessment results, presented herein, are quantifying hypothetical resources, according to the classification scheme of USGS (1980). The sampled material and the data collected during the MarMine research cruise will initiate the definition of inferred resources along the AMOR. This will, in the future, also be used to update the models presented in this book. It will, however, take many more research cruises before the AMOR deposits can be quantified as indicated or measured resources, and considerable work on all the modifying factors must be performed to render it possible to define any reserves.

Worldwide, the Solwara 1 Project inside the national jurisdiction of Papua New Guinea is the project that has come farthest in defining and classifying any SMS mineral resources on the deep ocean floor. The project owner, Nautilus Minerals Ltd, have published a report documenting extensive mapping and mineral classification according to NI43-101 (Lipton et al., 2012).

9.2 Resource potential

The results presented in this book were developed to review the knowledge about seabed mineral resources in Norway, with a special focus on possible massive sulphide mineralisation along the Mid-Atlantic Ridge. The work was carried out by NTNU and supported by Nordic Ocean Resources AS and Statoil ASA (now Equinor ASA).

The main area of interest was the extended Norwegian Continental Shelf, on the northern part of the Mid-Atlantic Ridge, an area with volcanic activity that has created the islands of Jan Mayen and Iceland. In the volcanic areas, boiling seawater extracts minerals and metals from the oceanic crust. When this comes into contact with cold water, gold, silver, copper and zinc are precipitated to a varying degree, dependent on the geological settings.

Analyses and interpretations of seabed topography, structures and geology have been carried out in order to disclose relevant areas for the formation of massive sulphides deposits on the ocean floor. Based on these data, a probabilistic mineral resource assessment was done using the same methodological framework as is used for oil and gas resources, by the Norwegian Petroleum Directorate and oil and gas companies worldwide. In the summer of 2016, NTNU together with 13 industry partners and funded by the Norwegian Research Council through the MarMine Project, carried out fieldwork in the area to collect samples from known sea floor massive sulphide deposits. The collected samples will, in the coming years, be used to update the input into the probabilistic analysis and thereby update the estimates. Work along these lines has been initiated (Juliani & Ellefmo, 2018a, 2018b and 2018c) and will continue in the years to come.

Information on the general locations and amounts of undiscovered seabed massive sulphide resources within the Norwegian Economic Exclusive Zone will be increasingly important to exploration and resource managers, environmental planners, economists and policymakers, as we possibly progress towards a sustainable extraction of these resources.

This book contains the results of probabilistic estimates of the amounts of metal of copper (Cu), zinc (Zn), silver (Ag), and gold (Au) in known and undiscovered seabed massive sulphide vent fields on the Mid-Atlantic Ridge, inside the Norwegian jurisdiction on the extended Norwegian Continental Shelf. The play analysis methodology that was used to make these estimates allows for an explicit expression of the estimates of undiscovered resources and their associated uncertainty in a form that is useful to decision makers. A combination of the probability distributions of the estimated number of undiscovered vent fields, the grades and the tonnages, was used to obtain the probability distributions for undiscovered metals. However, populating the input distributions consistently with robust data is not trivial. One has to think outside the box. It is essential to connect and combine the dots and pieces of data and information available. The grade data used as a basis in this study is based on global datasets, but all from mid-ocean ridge settings. Some data presented by Pedersen et al. (2010b) and data obtained from samples collected during the MarMine research cruise, show grade values lower than the grade data used in this play analysis. This data is scanty and based on surface samples only, and its significance relative to the much larger datasets presented by Cherkashev et al. (2013) and Hannington (2013a and 2013b) should not be emphasised.

A framework for the analysis has been established in this project. While we believe this framework is solid, the input parameters and data can always be improved, and the effects of improvements can be quantified immediately, especially in terms of reducing uncertainty. One issue that could be discussed is, of course, the use of a dependency factor quantifying the correlation between and inside plays in the aggregation of the plays. It goes almost without saying that it is wrong not to use one, but the real question lies in what the dependency factor should be. New ways of collecting and analysing data must be developed in order to quantify this parameter more precisely. Another issue not worked on here is to differentiate between the characteristics of the different permissive tracts. Given more local variability and coincidence of both geodynamical and morphostructural indicators, it would be possible to assign different properties to different permissive tracts. Further, a third issue is the possible correlation between, for example, the deposit grade and the deposit tonnage. Based on ancient deposits found and studied onshore, no clear correlations can be identified between these two parameters. However, large high-grade deposits are very rare. The effects of incorporating such trends or correlations are beyond the scope of this publication, but it would be interesting to look into the effects and how they could be incorporated into the play analysis.

There are currently five surveyed and a number of inferred seabed massive sulphide vent fields on the Norwegian Mid-Atlantic Ridge. To define favourable zones for vent fields, morphostructural and geodynamic analysis of bathymetric data was used according to the criteria established by the Russian Institute for Geology and Mineral Resources of the Ocean (VNIIOkeangeologia). In addition to the inferred vent fields, concealed vent fields may exist with less bathymetric expression than the ones inferred. This has been shown by internationally-acknowledged researchers who have published results that the assessment results obtained in this study have been validated against (Cathles, 2011a, 2011b; Hannington et al., 2011b; Hannington 2011, 2013b; Singer, 2014)

The total of inferred and postulated undiscovered metal (hypothetical resource) amounts to an expected endowment of 6.9 million metric tons of copper metal, in addition to zinc (7.1 million metric tons), gold (175 metric tons), and silver (10524 metric tons). Due to the lack of detailed data along this frontier exploration area, a great uncertainty is attached to these results with a huge upside potential. There is accordingly a 5% probability of having metal resources of more than or equal to 21 million metric tons of copper metal, 22 million metric tons of zinc, 597 metric tons of gold, and 34000 metric tons of silver. This corresponds to a gross rock value for the seabed minerals within Norwegian jurisdiction of NOK 600 billion. The estimation further indicates a possible potential of more than NOK 2000 billion. These values are, of course, very much dependent on exchange rates, because the commodities assessed here are traded on industrial metals markets like the London Metal Exchange and thereby traded in US dollars. The commodity prices are also volatile and highly dependent on trading conditions and states of the market. Further, the gross rock values are the 'value of the mineral resource in the ground' and the uncertainty is significant. The values do not take unknown cost of extraction, trans-shipment and processing and their uncertainties into account. Much must be done and investigated before the uncertainty is reduced and a decision can be made regarding a potential exploitation of these resources.

Nevertheless, due to the high potential indicated by this study, the Norwegian Government included marine minerals in their official minerals strategy in 2013, and in 2017, the Norwegian Government presented a proposal for a new Act on Offshore Mineral Exploration (Prop. 106 L (2017–2018) *Lov om mineralvirksomhet på kontinentalsokkelen (havbunns-mineralloven)*). It states that the Norwegian State is the owner of minerals on the Norwegian Continental Shelf and the bill introduces a new regime for private mineral exploration and extraction inspired by the Petroleum Act. The new law came into effect on 1 July 2019.

Further exploration in the years to come will hopefully confirm the existence of the inferred resources, as well as the costs and viability of their development, and thereby reduce the great uncertainty that at the present time is related to their tonnages and gross value. It is our strong recommendation that the Norwegian authorities provide the necessary funding to map the entire ridge, so that the uncertainty can be reduced and the viability of marine minerals as a future mining resource can be properly evaluated. This includes finalising a predictable and efficient administrative system, and encouraging and rendering it possible for universities and research institutions to do the necessary research.

About the Authors

Steinar L. Ellefmo is Associate Professor at the Department of Geoscience and Petroleum, Norwegian University of Science and Technology (NTNU). His research and teaching interests are within the broad field of mineral resource management and mining engineering, specifically geostatistics, 3D ore body modelling, mine planning and mineral resource potential assessment. Ellefmo is a member of the Australasian Institute of Mining and Metallurgy (AusIMM) and the Society of Economic Geologists (SEG).

Fredrik Søreide holds a PhD in Marine Engineering from the Norwegian University of Science and Technology (NTNU). He is currently Adjunct Professor at the Department of Marine Technology at NTNU focusing on underwater technology and marine science. He has been involved in marine minerals research and ocean mining activities for the past 20 years and is currently mainly engaged in the JPIO project, Ecological Aspects of Deep-sea Mining. Søreide is the author of three other books.

Georgy Cherkashov is Deputy Director of the Institute for Geology and Mineral Resources of the Ocean (VNIIOkeangeologia, St. Petersburg, Russia). He holds a Dr. Sci. for research of seafloor massive sulphide deposits of the Mid-Atlantic Ridge (2004). Cherkashov is a member of the Legal and Technical Commission of the International Seabed Authority (since 2012) and Professor of Marine Geology at St. Petersburg State University (since 2005).

Cyril Juliani received his PhD from the Norwegian University of Science and Technology (NTNU), Department of Geoscience and Petroleum. His dissertation focused on quantifying undiscovered seabed mineral resources and evaluating the exploration risk for deep-sea mineral prospects. Juliani is currently involved in the NTNU oceans pilot program on deep-sea mining, a research project focusing on the different value chains of seabed mineral extraction and exploration.

Krishna Kanta Panthi is Professor of Geological Engineering at the Department of Geoscience and Petroleum, Norwegian University of Science and Technology (NTNU). He holds a PhD (Dr.Ing.) in rock engineering, MSc in hydropower development and MSc in tunnelling. He has over 27 years of experience in research, design and implementation of rock engineering, tunnelling, hydropower, mining and slope engineering projects. Panthi is the author of over 70 scientific articles and book chapters published in internationally recognized journals, publication canals and conference proceedings.

Sergey Petukhov heads the analytical group Department of Geology and Mineral Resources of the Ocean, VNIIOkeangeologia. Petukhov holds a PhD in mining engineering and is an expert in data management and modelling of geodynamic processes.

Irina Poroshina is a senior scientist at VNIIOkeangeologia. She is an expert in geomorphology and structural analysis of mid-ocean ridges.

Richard Sinding-Larsen has been Professor of Resource Geology at the Norwegian University of Science and Technology (NTNU) since 1980, Emeritus since 2012. His research focuses on the estimation of undiscovered mineral resources, with particular attention to the process of discovering metals and petroleum. He has a long-standing interest in deposit modelling and pioneered probabilistic assessment of Caledonian massive sulphide resources and initiated the development of methodology and tools (GeoX, now owned by Schlumberger) used as decision support for risk, resource and economic evaluation of exploration. Sinding-Larsen's current research interest is in how to appraise and quantify resource uncertainties within mineral and petroleum systems whose components are logically related in complex ways – such as sulphides linked to volcanic spreading ridges beneath the oceans and petroleum systems.

Ben Snook achieved a PhD in granitic petrology after working on a variety of exploration geology projects in South America. He has continued to pursue research in Norwegian mineral deposits, specialising in the characterisation of seafloor massive sulphide ores and in automated mineralogy analytical systems.

References

Aleksandrov, P.A., Anikeeva, L.I., Andreev, S.I., & Petukhov, S.I. (2009). Thalassochemistry of the ocean ore genesis (in Russian). St. Petersburg: Federal State Unitary Enterprise 'VNIIOkeangeologia'.

Amundsen, I.M.H., Blinova, M., Hjelstuen, B.O., Mjelde, R., & Haflidason, H. (2011). The Cenozoic western Svalbard margin: Sediment geometry and sedimentary processes in an area of ultraslow oceanic spreading. *Marine Geophysical Research.* https://doi.org/10.1007/s11001-011-9127-z

Barriga, F.J.A.S., Carvalho, D. & Ribeiro, A. (1997). Introduction to the Iberian Pyrite Belt. In F.J.A.S. Barriga & D. Carvalho (Eds.), *Geology and VMS deposits of the Iberian Pyrite Belt guidebook series.* Volume 27. Society of Economic Geologists. https://doi.org/10.5382/GB.27

Barriga, F.J.A.S, Relvas, J.M.R.S, Santos, R., & Pascoal, A. (2013). Estimating and finding seafloor and sub-seafloor sulfide mineralization: Optimists versus pessimists? Conference report, Recent Developments in Atlantic Seabed Minerals Exploration and Other Topics. 21-28 October 2013. Rio de Janeiro and Porto de Galinhas, Brazil: Underwater Mining Institute.

Baker, E.T., & German, C.R. (2004). On the global distribution of mid-ocean ridge hydrothermal vent-fields. *Am Geophys Union Geophys Monogr, 148,* 245–66.

Baker, E.T. (2017). Exploring the ocean for hydrothermal venting: New techniques, new discoveries, new insights. *Ore Geology Reviews.* http://dx.doi.org/10.1016/j.oregeorev.2017.02.006

Baumberger, T., Frueh-Green, G.L., Pedersen, R., Thorseth, I.H., Lilley, M.D., & Moeller, K. (2010). Loki's Castle: A sediment-influenced hydrothermal vent field at the ultra-slow spreading Arctic Mid-Ocean Ridge. American Geophysical Union, Fall Meeting 2010, abstract #OS21A-1481.

Beaulieu, S.E. (2015). InterRidge global database of active submarine hydrothermal vent fields. Prepared for InterRidge, Version 3.3. Version 3.4 retrieved 03.01.2017 from http://vents-data.interridge.org

Becker, J.J., Sandwell, D.T., Smith, W.H.F., Braud, J., Binder, B., Depner, J., Fabre, D., Factor, J., Ingalls, S., Kim, S-H., Ladner, R., Marks, K., Nelson, S., Pharaoh, A., Trimmer, R., Von Rosenberg, J., Wallace, G. & Weatherall, P. (2009). Global Bathymetry and Elevation Data at 30 Arc Seconds Resolution: SRTM30_PLUS. *Marine Geodesy,* 32(4), 355–371. http://dx.doi.org/10.1080/01490410903297766. Data retrieved from http://www.pacioos.hawaii.edu/metadata/srtm30plus_v11_bathy.html

Braathen, A., & Gabrielsen, R.H. (2000). Bruddsoner i fjell – oppbygning og definisjoner. *Gråsteinen 7.* NGU.

Brekke, H. (2017). Framtidens mineraler. *Geo365*. Retrieved 03.06.2017 from
https://www.geo365.no/bergindustri/framtidens-mineraler/

Bruvoll, V., Breivik, A.J., Mjelde, R., & Pedersen, R.B. (2009). Burial of the Mohn-
Knipovich seafloor spreading ridge by the Bear Island Fan: Time constraints on
tectonic evolution from seismic stratigraphy. *Tectonics*, 28, TC4001.
https://doi.org/10.1029/2008TC002396

Cathles, L. M. (2011a). What processes at mid-ocean ridges tell us about
volcanogenic massive sulfide deposits. *Mineralium Deposita*, 46(5), 639-657.
https://doi.org/10.1007/s00126-010-0292-9

Cathles, L. M. (2011b). Reply to comment by Hannington: 'Are black smokers copper
plating the ocean floor?'. *Mineralium Deposita*, 46(5), 665–669.

Cathles, L. M. (2013). Future Rx: Optimism, preparation, acceptance of risk. In
G.R.T. Jenkin, P.A.J. Lusty, I. Mcdonald, M.P. Smith, A.J. Boyce & J.J. Wilkinson
(Eds.), *Ore deposits in an evolving earth*. London: Geological Society, London.
Retrieved from https://doi.org/10.1144/SP393.6

Charlou, J.L, Donval, J.P, Fouquet, Y, Jean-Baptiste P, & Holm, N. (2002).
Geochemistry of high H2 and CH4 vent fluids issuing from ultramafic rocks at
the Rainbow hydrothermal field (36j14VN, MAR). *Chemical Geology*, *191*,
345–359. http://dx.doi.org/10.1016/S0009-2541(02)00134-1

Chen, Y.J. (1992). Oceanic crustal thickness versus spreading rate. *Geophysical
Research Letters*, *19*, 753–756. https://doi.org/10.1029/92GL00161

Cherkashov, G.A., Ivanov, V.N., Bel'tenev, V.I., Lazareva, L.I., Rozhdestvenskayab,
I.I., Samovarov, M.L., Poroshina, I.M., Sergeev, M.B., Stepanova, T.V., Dobretsova,
I.G., & Kuznetsov, V.Y. (2013). Massive sulfide ores of the Northern Equatorial
Mid Atlantic Ridge. *Oceanology*, *53*(5) 607–619.

Cherkashov, G., Poroshina, I., Stepanova, T., Ivanov, V., Bel'tenev, V., Lazareva, L.,
Rozhdestvenskaya, I., Samovarov, M., Shilov, V., Glasby, G. P., Fouquet, Y, &
Kuznetsov, V. (2010). Seafloor massive sulfides from the northern Equatorial
Mid-Atlantic Ridge: New discoveries and perspectives. *Marine Georesources and
Geotechnology 28*, 222–239.

Cherkashov, G.A., Andreev, S., Petukhov, S., Poroshina, I., & Zayonchek, A. (2013).
Identification of SMS prospective areas at the mid-ocean ridges within the
Norwegian EEZ based on analysis of multibeam bathymetry data (Unpublished
report). Saint-Petersburg: VNII Okeangeologia.

Connelly, D.P., German, C.R., Naganuma, T., Pimenov, N., Egorov, A., & Dohsik,
H. (2002). Hydrothermal plume signals along the Knipovich Ridge. American
Geophysical Union, Fall Meeting 2002, abstract #T11A-1229.

Cook Islands Seabed Minerals Authority. (2018). Retrieved 05.11.2018 from
http://www.seabedmineralsauthority.gov.ck

Crane, K., & Solheim, A. (1995). Seafloor atlas of the Norwegian-Greenland Sea.
Tromsø: Norsk Polarinstitutt.

Crane, K., Doss, H., Vogt, P., Sunsvor, E., Cherkashov, G., Poroshina, I., & Joseph, D. (2001). The role of the Spitsbergen shear zone in determining morphology, segmentation and evolution of the Knipovich ridge. *Marine Geophysical Research, 22,* 153–205.

Cronan, D.S. (1992). *Marine minerals in exclusive economic zones.* London: Chapman & Hall.

Cronan, D.S. (Ed.) (1999). *Handbook of marine mineral deposits.* Boca Raton, FL: CRC Press.

Dauteuil, O., & Brun, J.-P. (1993). Oblique rifting in a slow-spreading ridge. *Nature, 361,* 145–148. https://doi.org/10.1038/361145a0

Dauteuil, O., & Brun, J.P. (1996). Deformation partitioning in a slow spreading ridge undergoing oblique extension: Mohns Ridge, Norwegian Sea. *Tectonics, 15,* 870–884.

DeMets, C., Gordon, R., Argus, D., & Stein, S. (1990). Current plate motion. *Geophysical Journal International, 101,* 425–478.

Douville, E., et al. (2002). The rainbow vent fluids (36°14 N, MAR): The influence of ultramafic rocks and phase separation on trace metal content in Mid-Atlantic Ridge hydrothermal fluids. *Chemical Geology, 184,* 37–48.

Ellefmo, S.L., & Søreide, F. (2014). *Marine minerals in Norwegian waters, a survey of knowledge and research needs.* Trondheim: NTNU.

Faleide, J.I., Solheim, A., Fiedler, A., Hjelstuen, B.O., Andersen, E.S., & and Vanneste, K. (1996). Late Cenozoic evolution of the western Barents Sea-Svalbard continental margin. *Global Planet Change, 12,* 53–74. https://doi.org/10.1016/092 -8181(95)00012-7

Fiedler, A., & Faleide, J.I. (1996). Cenozoic sedimentation along the southwestern Barents Sea margin in relation to uplift and erosion of the shelf. *Global and Planetary Change, 12,* 75–93. https://doi.org/10.1016/0921-8181(95)00013-5

Fouquet, Y., Wafik, A., Cambon, P., Mevel, C., Meyer, G., & Gente, P. (1993). Tectonic setting and mineralogical and geochemical zonation in the Snake Pit sulfide deposit (Mid-Atlantic Ridge at 23°N). *Economic Geology, 88,* 2018–2036.

Fouquet, Y., Charlou, J.L., Ondréas, H., Radford-Knoery, J., Donval, J.P., Douville, E., Apprioual, R., Cambon, P., Pellé, H., Landuré, J.Y., Normand, A., Ponsevera, E., German, C., Parson, L., Barriga, F., Costa, I., Relvas, J., & Ribeiro, A. (1997). Discovery and first submersible investigations on the Rainbow hydrothermal field on the MAR (36°14′ N). American Geophysical Union Fall Meeting, San Francisco. *Eos Transactions American Geophysical Union, 78*(46), 832.

Fouquet, Y., Cambon, P., Etoubleau, J., Charlou, J. L., Ondréas, H., Barriga, F. J. A. S. Cherkashov, G., Semkova, T., Poroshina, I. Bohn, M. Donval, J. P., Henry, K., Murphy, P., & Rouxel, O. (2010). Geodiversity of hydrothermal processes along the Mid-Atlantic Ridge and ultramafic-hosted mineralization: A new type of oceanic Cu-Zn-Co-Au volcanogenic massive sulfide deposit. In P.A. Rona, et al., *Diversity of hydrothermal systems on slow spreading ocean ridges* (pp. 321–368). Washington, DC: American Geophysical Union.

Fournier, M., & Petit, C. (2007). Oblique rifting at oceanic ridges: Relationship between spreading and stretching directions from earthquake focal mechanisms. *Journal of Structural Geology, 29*, 201–208. https://doi.org/10.1016/j.jsg.2006.07.017

Géli, L., Renard, V., & Rommevaux, C. (1994). Ocean crust formation processes at very slow spreading centers: A model for the Mohns Ridge, near 72°N, based on magnetic, gravity,and seismic data. *Journal of Geophysical Research, 99*, 2995–3013. https://doi.org/0148-0227/94/9 3JB-02966505.00

GeoX (2014). Schlumberger. Retrieved January 2014 from http://www.software.slb.com/lists/salesandmarketingdocuments/attachments/313/geox_prospect_asses sment_solution.pdf

German, C.R., Thurnherr, A.M., Knoery, J., Charloub, J.L., Jean-Baptiste, P., & Edmonds, H.N. (2010). Heat, volume and chemical fluxes from submarine venting: A synthesis of results from the Rainbow hydrothermal field, 36 degrees N MAR. *Deep-sea Research Part I, Oceanographic Research Papers, 57*(4) 518–527. Retrieved from http://archimer.ifremer.fr/doc/00003/11423/

Gràcia, E., Charloub, J.L,. Radford-Knoery, J., & Parson, L.M. (2000). Non-transform offsets along the Mid-Atlantic Ridge south of the Azores (38°N–34°N): Ultramafic exposures and hosting of hydrothermal vents. *Earth and Planetary Science Letters 177*(1–2), 89–103.

Granot, R., Abelson, M., Ron, H., Lusk, M.W., & Agnon, A. (2011). Direct evidence for dynamic magma supply fossilized in the lower oceanic crust of the Troodos ophiolite. *Geophysical. Research Letters, 38*, L16311. https://doi.org/10.1029/2011GL048220

Grindlay, N.R., Fox, P.J., & MacDonald, K.C. (1991). Second-order ridge axis discontinuities in the south Atlantic: Morphology, structure, and evolution. *Marine Geophysical Researches, 13*(1), 21–49.

Grindlay, N.R., Fox, P.J., & Vogt, P.R. (1992). Morphology and tectonics of the Mid-Atlantic Ridge (25°- 27°30′S) from Sea Beam and magnetic data. *Journal of Geophysical Research, 97*(B5), 6983–7010.

Hannington, M.D, Petersen, S., Herzig, P.M., & Jonasson, I.R. (2004). A global database of seafloor hydrothermal systems, including a digital database of geochemical analyses of seafloor polymetallic sulfides. Geological Survey of Canada, Open File, 4598, 2004. Retrieved from https://oceanrep.geomar.de/33243/1/Hannington.pdf

Hannington, M.D., de Ronde, C.E.J., & Petersen, S. (2005). Sea-floor tectonics and submarine hydrothermal systems. In J. Hedenquist et al., (Eds.), *100th anniversary volume of economic geology* (pp. 111–141). Littleton, CO: Society of Economic Geologists, Inc.

Hannington, M.D. & Monecke, T. (2009). Global exploration models for polymetallic sulphides in The Area. Assessment of the draft regulations on prospecting and exploration for polymetallic sulphides. *Marine Georesources & Geotechnology, 27*, 132–159.

Hannington, M., Jamieson, J., Monecke, T., & Petersen, S. (2010). Modern sea-floor massive sulfides and base metal resources: Toward an Estimate of Global SMS Potential Seafloor Massive Sulfide Deposits. *Society of Economic Geologists Special Publication, 15*, 317–338.

Hannington, M.D., Jamieson, J.W., Monecke, T., & Petersen, S. (2011a). Estimating the metal content of SMS deposits. (conference report, pp. 1–4). *OCEANS, IEEE*.

Hannington, M., Jamieson, J., Monecke, T., Petersen, S., & Beaulieu, S. (2011b). The abundance of seafloor massive sulfide deposits. *Geology. 39*(12), 1155–1158.

Hannington, M. D., Jamieson, J. W, Monecke, T. and Petersen, S. (2013a). Towards an estimate of global SMS potential. Kiel, Germany: Futureocean Society-Integrated School of Ocean Sciences (ISOS), Christian-Albrechts-Universität. Retrieved from http://fileserver.futureocean.org/forschung/r3/Hannington_SMS_resource_potential.pdf

Hannington, M.D. (2013b). The role of black smokers in the Cu mass balance of the oceanic crust. *Earth and Planetary Science Letters, 374*, pp. 215–226. http://doi.org/10.1016/j.epsl.2013.06.004

Harrison, J.P. & Hudson, J.A. (2000). *Engineering rock mechanics: Part 2. Illustrative worked examples*. Oxford: Pergamon.

Havbunnsmineralloven (2019). Lov om mineralvirksomhet på kontinentalsokkelen (LOV-2019-03-22-7). Retrieved from https://lovdata.no/dokument/NL/lov/2019-03-22-7

Ingri, J. (1985). Geochemistry of Ferromanganese Concretions in the Barents Sea. *Marine Geology, 67*, 101–119.

ISA (2018). Website of the International Seabed Authority. Retrieved 18.03.2018 from https://www.isa.org.jm/deep-seabed-minerals-contractors/overview

Johansen, S.E., Panzner, M., Mittet, R., Amundsen, H.E.F., Lim, A., Vik, E., Landrø, M., & Arntsen, B. (2019). Deep electrical imaging of the ultraslow-spreading Mohns Ridge. *Nature, 567*, 379–383. https://doi.org/10.1038/s41586-019-1010-0

JORC (2012). The JORC Code. Retrieved 18.03.2018 from http://www.jorc.org/docs/JORC_code_2012.pdf

Jowitt, S.M., Jenkin, G.R.T., Coogan, L.A., & Naden, J. (2012). Quantifying the release of base metals from source rocks for volcanogenic massive. *Journal of Geochemical Exploration, 1*(18), 47–59.

Juliani, C.J., & Ellefmo, S.L. (2018a). Resource assessment of undiscovered seafloor massive sulfide deposits on an Arctic mid-ocean ridge: Application of grade and tonnage models. *Ore Geology Reviews, 102*.

Juliani, C.J., & Ellefmo, S.L. (2018b). Multi-scale quantitative risk analysis of seabed minerals: Principles and application to seafloor massive sulfide prospects. *Natural Resources Research*. https://doi.org/10.1007/s11053-018-9427-y

Juliani, C.J., & Ellefmo, S.L. (2018c). Probabilistic estimates of permissive areas for undiscovered seafloor massive sulfide deposits on an Arctic Mid-Ocean Ridge. *Ore Geology Reviews, 95.*

Kandilarov, A., Landa, H., Mjelde, R., Pedersen, R.B., Okino, K., & Murai, Y. (2010). Crustal structure of the ultra-slow spreading Knipovich Ridge, North Atlantic, along a presumed ridge segment center. *Marine Geophysical Researches, 31*(3), 173–195. https://doi.org/10.1007/s11001-010-9095-8

Kleinrock, M.C., & Humphris, S.E. (1996). Structural control on sea-floor hydrothermal activity at the TAG active mound. *Nature, 382,*149–53.

Klingelhöfer, F., Géli, L., Matias, L., Steinsland, N., & Mohr, J. (2000). Crustal structure of a super-slow spreading centre: A seismic refraction study of Mohns Ridge. *Geophysical Journal International, 14*(2), 509–526.

Krasnov, S.G., Cherkashev, G.A., Stepanova, T.V., Batuyev, B.N., Krotov, A.G., Malin, B.V., Maslov, M.N., Markov, V.F., Poroshina, I.M., Samovarov, M.S., Ashadze, A.M., Lazareva, L.I., & Ermo-layev, I.K. (1995). Detailed geological studies of hydrothermal fields in the North Atlantic. In L.M. Parson, C.L. Walker & D.R. Dixon (Eds.), *Hydrothermal vents and processes* (pp. 43–64). Geological Society Special Publication, vol. 87.

Krysiński, L., Grad, M., & Mjelde, R. (2013). Seismic and density structure of the lithosphere–asthenosphere system along transect Knipovich Ridge–Spitsbergen–Barents Sea– geological and petrophysical implications. *Polar Polish Research, 34*(2), 111–138.

Lipton, I. (2012). Mineral Resource Estimate. Solwara Project, Bismarck Sea, PNG (report written for Nautilus Minerals).

Ludvigsen, M., Aasly, K., Ellefmo, S.L., Hilário, A., Ramirez-Llodra, E., Søreide, F. X., Falcon-Suarez, I., Juliani, C.J., Kieswetter, A., Lim, A., Malmquist, C., Nornes, S.M., Reimers, H., Paulsen, E., & Sture, Ø. (2016). NTNU Cruise reports 2016 no 1 - MarMine cruise report Arctic Mid-Ocean Ridge 15.08.2016–05.09.2016. Trondheim: NTNU.

Ludvigsen, M., Søreide, F., Aasly, K., Ellefmo, S.L., Zylstra, M., & Pardey, M. (2017). ROV based drilling for deep sea mining exploration. In conference proceedings, IEEE OCEANS 2017. Aberdeen: IEEE.

Mero, J.L. (1965). *Mineral resources of the sea.* First edition. Amersterdam/London/ New York: Elsevier Publishing Co.

Melekestseva, I.Y., Zaykov, V.V., Nimis, P., Tret'yakov, G.A. & Tessalina, S.G. (2013). Cu–(Ni–Co–Au) -bearing massive sulfide deposits associated with mafic-ultramafic rocks of the Main Urals Fault, South Urals: Geological structures, ore textural and mineralogical features, comparison with modern analogs. *Ore Geology Reviews, 52,* 18–36.

Mosar, J., Lewis, G., & Torsvik, T.H. (2002). North Atlantic sea-floor spreading rates: Implications for the Tertiary development of inversion structures of the Norwegian–Greenland Sea. *Journal of the Geological Society, 159,* 503–515.

Mosier, D.L., Singer, D.A., & Berger, V.I. (2007). Volcanogenic massive sulfide deposit density. s.l. USGS Scientific Investigations Report 2007–5082. Retrieved from http://pubs.usgs.gov/sir/2007/5082/sir2007-5082.pdf

Mosier, D.L., Berger, V.I. & Singer, D.A. (2009). Volcanogenic massive sulfide deposits of the world—database and grade and tonnage models. Retrieved from http://pubs.usgs.gov/of/2009/1034/

Mozgova, N.N., Bordaev, Y.S., Gablina, I.F., Cherkashev, G.A., & Stepanova, T.V. (2005). Mineral assemblages as indicators of the maturity of oceanic hydrothermal sulfide mounds. *Lithology and Mineral Resources, 40,* 293–319.

NI (2018). National Instrument 43-101. Retrieved 18.03.2018 from http://web.cim.org/standards/

NPD (2013). Norwegian Petroleum Directorate resource report 2013. Retrieved from https://www.npd.no/en/facts/publications/reports2/resource-report/archive-for-resource-reports/

NPD (2018a). Acquisition of seabed massive sulfide data in the Norwegian Sea. Qualifying documentation SMS data acquisition 2018. Retrieved 03.11.2018 from https://www.mercell.com/m/file/GetFile.ashx?id=80348247&version=0

NPD (2018b). Norwegian Petroleum Directorate maps deep sea mineral deposits. Retrieved 03.11.2018 from http://www.npd.no/en/news/News/2018/Norwegian-Petroleum-Directorate-maps-deep-sea-mineral-deposits/

NPD (2018c). The Norwegian Petroleum Directorate has found new deep sea mineral deposits. Retrieved 03.11.2018 from http://www.npd.no/en/news/News/2018/New-deep-sea-mineral-deposits/

NPD (2018d). Ministry of Petroleum and Energy responsible for mineral deposits on the shelf. Retrieved 29.12.2018 from http://www.npd.no/en/news/News/2017/Ministry-of-Petroleum-and-Energy-responsible-for-mineral-deposits-on-the-shelf-/

NPD (2019). Successful exploration for seabed minerals. Retrieved 10.09.2019 from https://www.npd.no/en/facts/news/general-news/2019/successful-exploration-for-seabed-minerals/

OED (2017). Lov om mineralvirksomhet på kontinentalsokkelen. Prop. 106 L (2017–2018). Draft legislation for the management of marine minerals on the Norwegian Continental Shelf. Retrieved from https://www.regjeringen.no/no/dokumenter/prop.-106-l-20172018/id2605252/

Okino, K., Curewitz, D., Asada, M., Tamak, i K., Vogt, P., & Crane, K. (2002). Preliminary analysis of the Knipovich Ridge segmentation: Influence of focused magmatism and ridge obliquity on an ultraslow spreading system. *Earth and Planetary Science Letters, 202,* 375–288.

Olsen, B.R, Økland, I.E, Thorseth, I.H., Pedersen, R.B., & Rapp, H.T. (2016). Environmental challenges related to offshore mining and gas hydrate extraction. Report M-532 2016. Trondheim: Miljødirektoratet.

Petukhov, S.I., & Aleksandrov P.A. (2004). Application of a method of geodynamic zoning for identification of fields of hydrothermal activity in the region of mid-Atlantic ridge (for MAR – 15 0 30´ – 17030´ N.). Minerals of the ocean – integrated strategies-2. Conference abstract, SPb, 2004, pp. 41–42.

Petukhov, S.I, Aleksandrov, P.A., & Andreev, S.I. (2010). Deformation model of hydrothermal sulfide ore fields for prediction of hydrothermal activity locations (for different areas of the Atlantic and Indian oceans). Minerals of the ocean – 5 and deep-sea minerals and mining-2. Conference abstract, SPb, 2010, pp. 82–83.

Petukhov, S.I, Aleksandrov, P.A., & Tao, C. (2012). Modeling of the environments favourable for hydrothermal activity using relief analysis (for South Atlantic Ridge). Minerals of the ocean – 6 and deep-sea minerals and mining-3. Conference abstract, SPb, 2012, pp. 136–137.

Pedersen, R.B, Rapp, H.P., Thorseth, I.H., Dilley, M.D., Barriga, F.J.A.S., Baumberger, T., Flesland, K., Fonseca, R., Früh-Green, G.L., & Jørgensen, S.L. (2010a). Discovery of a black smoker vent field and vent fauna at the Arctic Mid- Ocean Ridge. *Nature communication.* https://doi.org/10.1038/ncomms1124

Pedersen, R.B, Thorseth, I.H., Nygård, T.E., Dilley, M.D., & Kelley, D.S. (2010b). Hydrothermal activity at the arctic mid-ocean ridges. In P.A. Rona, et al., *Diversity of hydrothermal systems on slow spreading ocean ridges* (pp. 67–89). Washington, DC: American Geophysical Union.

Panthi, K.K. (2006). Analysis of engineering geological uncertainties related to tunneling in Himalayan rock mass conditions. (PhD thesis.) Trondheim: Norwegian University of Science and Technology.

Panthi, K.K. (2012). Evaluation of rock bursting phenomena in a tunnel in the Himalayas. *Bulletin of Engineering Geology and the Environment, 71*(4), 761–769.

Panthi, K.K. (2016). Introduction to engineering geology. Lecture notes on TGB5110 Engineering Geology and Tunneling BC. Norwegian University of Science and Technology (NTNU).

Prandtl, L. (1904). Über Flussigkeitsbewegungen bei sehr kleiner Reibung (Motion of fluids with very little viscosity). In A. Krauzer (Ed.), Proceeding of the Third International Mathematics Congress, Heidelberg, Germany (pp. 484–491). Publisher unknown. See English translation from 2001 in J. A. K. Ackroyd, B. P. Axcell & A. I. Ruban (Eds.), *Early developments of modern aerodynamics* (p. 77). Oxford: Butterworth-Heinemann.

Robb, L. (2004). *Introduction to ore-forming processes.* Oxford: Wiley-Blackwell.

Rona, P.A. & Scott, S.D. (1993). A special issue on sea-floor hydrothermal mineralization: New perspectives. *Economic Geology, 88,* 1935–75.

Rona, P.A. (2008). The changing vision of marine minerals. *Ore Geology Review, 33*, 618–66.

Rona, P.A., Devey, C.W., Dyment, J., & Murton, B.J. (2010). Diversity of hydrothermal systems on slow spreading ocean ridges. Washington, DC: American Geophysical Union.

Rowland, S.M., & Duebendorfer, E.M. (1994). *Structural analysis and synthesis. A laboratory course in structural geology,* 2nd Edition. Malden, MA/Oxford/ Melbourne/Berlin: Blackwell Scientific Publications.

Ryan, W.B.F., Carbotte, S.M., Coplan, J.O., O'Hara, S., Melkonian, A., Arko, R., Weissel, R.A., Ferrini, V., Goodwillie, A., Nitsche, F., Bonczkowski, J., & Zemsky, R. (2009). Global multi-resolution topography synthesis, *Geochemistry. Geophysics. Geosystems, 10.* https://doi.org/10.1029/2008GC002332

Schmidt, K., Koschinsky, A., Garbe-Schönberg, D., de Carvalho, L. M., Seifert, R. (2007). Geochemistry of hydrothermal fluids from the ultramafic-hosted Logatchev hydrothermal field, 15°N on the Mid-Atlantic Ridge: Temporal and spatial investigation. *Chemical Geology, 242*(1–2), 1–21. Retrieved from http://dx.doi.org/10.1016/j.chemgeo.2007.01.023

Scott, S. (2004). Proposed exploration and mining technologies for polymetallic sulphides. In proceedings from workshop for the establishment of environmental baselines at deep seafloor cobalt-rich crusts and deep seabed polymetallic sulphide mine sites in the area for the purpose of evaluating the likely effects of exploration and exploitation on the environment. International Seabed Authority, Kingston, Jamaica, 6-10 September 2004. Retrieved from https://ran-s3. s3.amazonaws.com/isa.org.jm/s3fs-public/documents/EN/Workshops/2004/ Proceedings-ae.pdf

Sempéré, J.C., Lin, J., Brown, H.S. Schouten, H., & Purdy, G.M. (1993). Segmentation and morphotectonic variations along a slow-spreading center: The Mid-Atlantic Ridge (24°00′ N- 30°40′ N). *Marine Geophysical Researches, 15*(3), 153–200.

Seyfried Jr., W.E., et al. (2011). Vent fluid chemistry of the Rainbow hydrothermal system (36 N, MAR): Phase equilibria and in situ pH controls on subseafloor alteration processes. *Geochimica et Cosmochimica Acta, 75,* 1574–1593.

Sinding-Larsen, R. & Ellefmo, S. (2014). Seabed massive sulphide resource assessment of undiscovered potentially recoverable copper, zinc, silver, and gold related to hydrothermal vent fields on the Mid-Atlantic ridge within the Norwegian Economic Exclusive Zone. In S.L. Ellefmo & F. Søreide, *Marine minerals in Norwegian waters, a survey of knowledge and research needs.* Trondheim: NTNU.

Sinding-Larsen, R. & Vokes, F. (1978). The use of deposit modeling in the assessment of potential resources as exemplified by Caledonian stratabound sulfide deposits. *Mathematical Geology, 10*(5), 565–579. https://doi.org/10.1007/BF02461986

Singer, D. (1986). Descriptive model of Cyprus massive sulfide. In D.P. Cox & D.A. Singer, *Mineral deposit models, U.S. Geological Survey Bulletin, 1693* (p. 131). Washington, DC: U.S. Geological Survey.

Singer, D.A. (2014). Base and precious metal resources in seafloor massive sulfide deposits. *Ore Geology Reviews, 59*, 66–72. http://doi.org/10.1016/j.oregeorev.2013.11.008

Singer, D.A., & Mosier, D.L. (1986). Grade and tonnage model of Cyprus massive sulfide. In D.P. Cox & D.A. Singer, *Mineral deposit models, U.S. Geological Survey Bulletin, 1693* (pp. 190–197). Washington, DC: U.S. Geological Survey.

Singer, D.A., Menzie, W.D. & Sutphin, D.M. (2001). Mineral deposit density—an update. In K.J. Schulz (Ed.), Contributions to global mineral resource assessment research: U.S. Geological Survey professional paper 1640-A (pp. A1-A13). Retrieved from https://pubs.usgs.gov/pp/p1640a/

Singer, D.A., & Menzie, W.D. (2008). Map scale effects on estimating the number of undiscovered mineral deposits. *Natural Resources Research, 17*(2), 79–86.

Singer, D.A., & Menzie, W.D. (2010). *Quantitative mineral resource assessments* (p. 232). Oxford: Oxford University Press.

Smith, D.K., Escartín, J., Schouten, H., & Cann, J.R. (2008). Fault rotation and core complex formation: Significant processes in seafloor formation at slow-spreading mid-ocean ridges (Mid-Atlantic Ridge, 13°–15°N). *Geochemistry, Geophysics, Geosystems, 9.* https://doi.org/10.1029/2007GC001699

Sundvor, E. (1997). Joint U.S., Russian and Norwegian research at Knipovich Ridge, Norwegian-Greenland Sea. *InterRidge News, 6*(1), 17.

Søreide, F., Lund, T., & Markussen, J.M. (2001). Deep ocean mining reconsidered – a study of the manganese nodule deposits in the Cook Islands, The 4th ISOPE Ocean Mining Symposium (pp. 88–93). Proceedings of the Fourth (2001) ISOPE Ocean Mining Symposium, Szczecin, Poland, International Society of Offshore and Polar Engineers.

Søreide, F., Ludvigsen, M., & Williamson, M. (2017). Low-cost drilling for deep-sea minerals exploration. *Sea Technology, 58*(8), 10–14.

Talwani, M., & Eldholm, O. (1977). Evolution of the Norwegian-Greenland Sea. *Geological Society of America Bulletin, 88*, 969–999.

Tamaki, K., & Cherkashev, G. (2000). Knipovich-2000 onboard cruise report, Ocean Res. Inst., Univ. of Tokyo, Tokyo. Retrieved from https://agupubs.onlinelibrary. wiley.com/doi/pdf/10.1029/2007GC001652

Tamaki, K., et al. (38 authors) (2001). Japan-Russia cooperation at the Knipovich Ridge in the Arctic Sea. *InterRidge News, 10*(1), 48–51.

Thermo Fisher (2018). Title and/or description. Retrieved 02.03.2018 from https://www.thermofisher.com/order/catalog/product/XL3TGOLDDPLUS

Torsvik, T.H., Mosar, J., & Eide, E.A. (2001). Cretaceous-Tertiary geodynamics: A North Atlantic exercise. *Geophysical Journal International, 146*, 850–866.

USGS (2009). Appendix C: A resource/reserve classification for minerals. In *Mineral Commodity Summaries* (pp.191–193). Retrieved 20.05.2019 from http://minerals.usgs.gov/minerals/pubs/mcs/2009/mcsapp2009.pdf

USGS (1980). Principles of a Resource/Reserve Classification for Minerals. *US Geological Survey Circular 831*. U.S. Bureau of Mines and U.S. Geological Survey. https://doi.org/10.3133/cir831

Vogt, P.R. (1974). Volcano spacing fractures and thickness from elastic flexure theory. *Geophysical Research Letters, 5*, 977–980.

Vogt, P.R., Perry, R.K., Feden, R.H., Fleming, H.S., & Cherkis, N.Z. (1981). The Greenland-Norwegian Sea and Iceland environment: Geology and geophysics. In B.G. Hurdle (Ed.), *The Arctic Ocean*, Volume 5. (pp. 493–598). New York: Springer-Verlag.

Vogt, P.R., Kovacs, L.C., Bernero, C., & Srivastava, S.P. (1982). Asymmetrical geophysical signatures in the Greenland-Norwegian and Southern Labrador Seas and the Eurasia Basin. *Tectonophysics, 89*, 95–150.

Vogt, P.R. (1986). The present plate boundary configuration. In P.R. Vogt & B.E. Tucholke (Eds.), *The geology of North America* (pp.189–204). Boulder, CO: Geology Society of America, Western North Atlantic Region.

Vorren, T.O., Laberg, J.S., Blaume, F., Dowdeswell, J.A., Kenyon, N.H., Mienert, J., Rumohr, J., & Werner, F. (1998). The Norwegian-Greenland Sea continental margins: Morphology and late quaternary sedimentary processes and environment. *Quaternary Science Reviews, 17*, 273–302.

Zayonchek, A.V., Sokolov, S.Y., Mazarovich, A.O., Ermakov, A.V., Razumovskii, A.A., Akhmedzyanov, V.R., Barantsev, A.A., Zhuravko, N.S., Moroz, E.A., Sukhikh, E.A., Fedorov, M.M., & Yampol'skii, K.P. (2011). Structure of the transition zone between Hovgaard Ridge and Spitsbergen Plateau according to the data obtained during cruise 27 of the RV Academik Nikolai Strakhov. https://doi.org/10.1134/S1028334X11080101

Zonenshain, L.P., Kuzmin, M.I., Lisitsin, Bogdanov, Y.A., & Baranov, B.V. (1989). Tectonics of the Mid-Atlantic rift valley between the TAG and MARK areas (260-240N): Evidence for vertical tectonism. *Tectonophysics, 159*, 1–23.